Valerii A. Isidorov

Organic Chemistry of the Earth's Atmosphere

Translated by E. A. Koroleva

With 37 Figures

Springer-Verlag Berlin Heidelberg NewYork
London Paris Tokyo HongKong

Valerii A. Isidorov
Department of Chemistry
Leningrad University
Leningrad 198 904, USSR

ISBN-13: 978-3-642-75096-0   e-ISBN-13: 978-3-642-75094-6
DOI: 10.1007/978-3-642-75094-6

Library of Congress Cataloging-in-Publication Data
Isidorov, V. A. (Valerii Alekseevich)
[Organicheskaia khimiia atmosfery. English]
Organic chemistry of the Earth's atmosphere / V. A. Isidorov;
translated by E. A. Koroleva. p. cm.
Translation of: Organicheskaia khimiia atmosfery.
Includes bibliographical references.

1. Atmospheric chemistry. 2. Chemistry, Physical organic.
3. Environmental chemistry. I. Title.
QC879.6.I7713 1990
551.5'11—dc20

© Springer-Verlag Berlin Heidelberg 1990
Softcover reprint of the hardcover 1st edition 1990

Typesetting: Thomson Press, India

2151/3020-543 210 – Printed on acid-free paper

Foreword by the Editor of the Russian Edition

The term "organic chemistry of the atmosphere" appears for the first time as a book title. Monographs on the chemistry of the atmosphere have already been published but they are concerned almost exclusively with its inorganic components. However, in the 1970s, owing to the development of chromatography-mass spectrometry techniques and concentration procedures, the information on organic components at atmospheric air accumulated so rapidly and the significance of organic compounds in the processes occurring in the Earth's atmosphere became so evident that there was an urgent need for the systematization and generalization of numerous data on these problems scattered in specialized journals dealing with quite different subjects.

It may be affirmed that in the 1970s the basis of organic chemistry of the atmosphere as an independent branch of science was laid. This branch covers complex problems involving the chemistry of organic compounds present in atmospheric air, their sources the paths of their further transformations, their role in meteorological processes, in formation and evolution of the climate, and the ecologic conditions governing the existence of humanity. The problems of organic chemistry of the atmosphere also include the questions related to the procedures of detection, identification and quantitative determination of negligible amounts of organic substances, the overall content and balance of which in the atmosphere attain astronomic figures greatly exceeding the scales of their industrial production. The problems and the limits of each new branch of science cannot be considered as being quite definitive. Consequently, the content of this book reflects some views of the author who has been working in this field for more than 15 years and has obtained many important results which, in my opinion, cover problems not just of local and particular importance.

The accentuation of some problems and some opinions of the author may be regarded as debatable, which is quite natural when we consider a new direction formed as an interdisciplinary field of several branches of science: geophysics, chemical ecology, and organic chemistry. There can be no doubt, however, that this book will be useful to the workers in these branches of science as well as to engineers and other specialists in the field of environmental protection.

1984

B. V. Ioffe
Professor of Leningrad University

Preface

The two past decades have been characterized by a drastically increasing interest in the chemical processes occurring in the Earth's atmosphere. The necessity for the solution of practical problems of counteracting the negative consequences of the ever increasing industrial and agricultural activity of man has stimulated both applied and fundamental investigations in this field. The accumulation of experimental observations and their interpretation have gradually led to the conclusion that a number of minor gaseous components, the total concentration of which does not even attain 1 part in 10 000 000 of the atmosphere, play a decisive part in such global processes as the formation of the radiation regime of the atmosphere and the Earth's surface, the general circulation of the atmosphere, and the climate of the planet. Suddenly it was realized that humanity has appeared and exists at the bottom of a gigantic natural chemical reactor, the Earth's atmosphere, and the conditions of our existence depend on its activity. Little by little, the finest threads have begun to the established relating the composition of microcomponents and their transformation to macro-scale planetary processes which until recently have been the fields of study of meteorology, climatology, and other geophysical sciences. As a result, it has become clear that all these macroprocesses should be studied at a molecular level.

A certain amount of progress in this field has already been attained owing to joint efforts of scientists in many countries. The results of computer simulation of changes in the planet's climate in connection with the evolution of the chemical composition of the atmosphere have been particularly impressive. However, some of the attempts which are made to represent the simulation results as predictions of changes in the radiation regime of the atmosphere, the climate, the state of the ozone layer of the stratosphere, etc, should be viewed with skepticism. The author is profoundly convinced that the present level of our knowledge of the composition of the atmosphere, the time-dimensional distribution of minor gaseous components and aerosols, their sources and sinks does not yet allow such predictions to be made. It should be borne in mind that the harm resulting from too optimistic and from too pessimistic conclusions of these predictions is approximately equal. A calm, sober, and impartial approach based on rigorous scientific facts and their comprehensive interpretation is needed.

I have attempted to carry out an initial generalization of the data accumulated during the past decades on the organic component of the Earth's atmosphere.

I consider this book to be a very modest contribution to the joint efforts of my colleagues in many countries who are anxious about the fate of Nature which is being more and more rapidly transformed into the "environment".

Leningrad, February 1990   V. A. Isidorov

Contents

1 Chemical Composition of the Earth's Atmosphere and Its Evolution

1.1 General Information About the Structure and Chemical Composition of the Atmosphere

The atmosphere is a gaseous envelope surrounding the Earth and revolving with it. Its mass is about 5.15×10^{15} t. Its upper boundary at which gases disperse into interplanetary space lies at an altitude of approximately 1000 km above sea level. The lower layer about 5.5 km thick contains half of the atmosphere mass, whereas the layer 40 km thick contains more than 99% of this mass.

High-altitude investigations which began exactly a hundred years ago have shown that the structure of the atmosphere is complex. The character of changes in temperature with increasing altitude indicates that several layers exist, and they are separated by narrow transition zones-pauses (Fig. 1.1). The lowest layer adjoining the Earth, the troposphere, is characterized by an average vertical temperature gradient of $6\,°\text{C km}^{-1}$. The height of the upper boundary of the troposphere varies from 8 km in high latitudes to 16 to 18 km above the equator. The troposphere is separated from the higher level, the stratosphere, by a

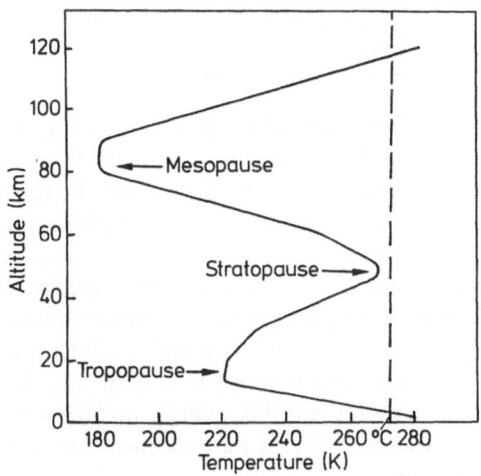

Fig. 1.1. Vertical distribution of temperature in the atmosphere

relatively narrow zone, the tropopause. The height of the tropopause in different latitudes does not vary monotonously, the tropopause has breaks and, as has been elucidated in recent years, forms folds. The latitudinal boundary of the break between the tropical and the polar tropopause also varies and is usually located between 35 and 50° latitude in each hemisphere. In the stratosphere, the temperature remains approximately constant up to an altitude of 25 km and then increases gradually to 200–220 K at the lower boundary of the stratopause (about 55 km). The mesosphere located higher is characterized by another temperature decrease to 190–180 K at an altitude of 80 km.

The thermosphere is located above the mesosphere and is separated from it by a transition layer. The kinetic temperature in the thermosphere increases with altitude almost regularly up to 1000–1500 K.

The variations in temperature in different layers of the atmosphere are caused by a change in the chemical composition of air in these layers. The Earth's atmosphere consists mainly of nitrogen and oxygen with small admixtures of other gases. The chemical composition of dry atmospheric air near the Earth's surface is given below:

	Volume fraction, %		Volume fraction, %
Nitrogen	78.084	Nitrogen monoxide	3.04×10^{-5}
Oxygen	20.9476	Xenon	8.7×10^{-6}
Argon	0.934	Sulphur dioxide	$\leqslant 7 \times 10^{-6}$
Carbon dioxide	3.45×10^{-2}	Ozone	$\leqslant 2 \times 10^{-6}$
			(in winter)
Neon	1.818×10^{-3}		$\leqslant 7 \times 10^{-6}$
Helium	5.24×10^{-4}		(in summer)
Methane	1.6×10^{-4}	Nitrogen dioxide	$\leqslant 2 \times 10^{-6}$
Cryptone	1.14×10^{-4}	Carbon oxide	5×10^{-5}
			-8×10^{-5}
Hydrogen	5×10^{-5}	Nitrogen oxide	$< 1 \times 10^{-4}$
		Amonia	$< 1 \times 10^{-4}$

On the whole the region below 90 km is characterized by intensive mixing and hence is relatively constant in composition. However, the concentrations of individual components vary over a comparatively wide range. One of the most important variables is the amount of water vapor present in the atmosphere. The content of water vapor rapidly decreases with altitude up to the tropopause. In the stratosphere, this content is very low (about 2×10^{-6}) and its dependence on altitude is very slight. The latitudinal gradient of water vapor concentration near the Earth's surface is also very pronounced: in the tropical regions this concentration attains 3%, whereas in the Antarctica it drops to 2×10^{-5}%. The water vapor is one of the main absorbers of the solar energy and thermal radiation of the Earth's surface. Hence, the decrease in its concentration with altitude results in a decrease in temperature.

The content of another variable component of the atmosphere, ozone, has a major effect on the thermal regime in the stratosphere. The air of the stratosphere is heated as a result of the exothermic decomposition of ozone molecules due to the effect of ultraviolet solar radiation. The greatest amounts of ozone are present in a layer located at an altitude from 20 to 30 km. In the mesosphere, the concentrations of ozone and water vapor are negligible and hence the temperature in this zone is lower than in the troposphere and the stratosphere. The increase in temperature in the thermosphere is caused by the absorption of a hard component of solar radiation by the molecules and atoms of oxygen and nitrogen. In this region, the strongest dependence of the chemical composition of air on altitude is observed: with increasing distance from the Earth's surface, the atmosphere is enriched with lighter gases as a result of gravitational separation. In a layer located at an altitude of 100 to 200 km, the major components are still nitrogen and oxygen, whereas above 600 km helium and hydrogen predominate.

1.2   Thermal Balance and Atmospheric Circulation

Solar radiation is the source of almost all energy on Earth. The solar constant characterizing the total energy flux (measured beyond the atmosphere) falling on 1 cm^2 of area perpendicular to the direction of the sun's rays per min is 8.2 J cm^{-2} min^{-1}. The average amount of radiation arriving in the upper layers of the atmosphere is about 1050 kJ cm^{-2} y^{-1}. One third of this radiation (350 kJ cm^{-2} y^{-1}) is reflected by the atmosphere and the Earth's surface into interplanetary space. The Earth's surface and the atmosphere absorb 450 and 250 kJ cm^{-2} y^{-1}, respectively.

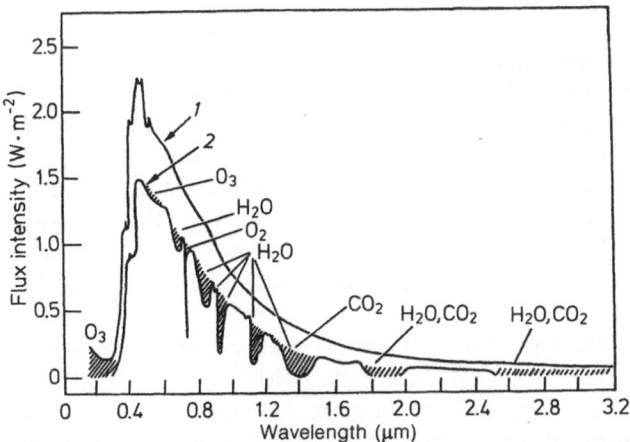

Fig. 1.2. Intensity of solar radiation *1*- beyond the atmosphere and *2*- at the sea level vs wavelength. *Shaded parts* correspond to the absorption of some components [1]

The major amount of solar energy is in the form of short-wavelength radiation with a maximum at the wavelength $0.47\,\mu m$. The dependence of distribution of radiation intensity on the wavelength is shown in Fig. 1.2. This figure indicates the regions of the absorption spectra of some air components. The hardest ultraviolet rays are retained in the stratosphere by the ozone screen. However, the main part of the solar energy passes through the upper layers and is partially absorbed in the troposphere by water vapor, carbon dioxide, oxygen, aerosols and dust particles. The radiation absorbed by the Earth's surface returns to the atmosphere in the form of the long-wave infrared radiation and is also consumed by water evaporation and the generation of convective turbulent air currents. The condensation of water vapor is accompanied by the evolution of heat consumed by heating the atmosphere. Only a small part of the long-wave radiation radiated back by the Earth's surface passes through the atmosphere and is dissipated in space. The main amount of this radiation is absorbed by the molecules of water and carbon dioxide, which leads to additional air heating. The thermal energy trapped by the atmosphere is again radiated back towards the Earth's surface, and as a result the greenhouse (atmospheric) effect is developed.

The amount of solar radiation arriving in the atmosphere depends on the angle of incidence of solar rays on the Earth's surface. The greatest amount of energy is absorbed in low latitudes. As a result, the atmosphere in different regions of the terrestrial globe is heated irregularly. Particularly great differences in temperature at the surface are observed between the polar and the equatiorial regions. This irregularity of heating is the main reason for the general circulation of the atmosphere which is a complex large-scale system of air currents over the terrestrial globe. Some of these currents are relatively stable, whereas the others constantly change their direction. The energy of moving air masses is consumed by friction but is replenished by solar radiation. The existence of this circulation smooths down the temperature gradient of the atmospheric air in different regions. The air currents also transport water vapor from the oceans to the continental regions and lead to the averaging of the composition of major air constituents within the homosphere up to an altitude of about 90 km. Tradewinds are also stable air currents. These winds blow in the low latitudes of both hemispheres and are directed from the substropical regions to the equator. On the equator the air is heated, ascends slowly and at high altitudes turns to the poles. As a result, circular motion of air is observed: it ascends in the equatorial region and descends near the subtropics. This motion is called the trade-wind circulation cell (Hadley cell). In the course of this circulation, the exchange of the components of atmospheric air between the troposphere and the stratosphere probably takes place (Fig. 1.3).

In some tropical regions, other stable air currents between the ocean and the continent, the monsoons, are observed. In middle latitudes, the currents from west to east predominate. They include large-scale vortices: cyclones and anticyclones. The formation of these vortices is usual for non-tropical regions. The cyclonic activity leads to the complex and variable character of both regional and general atmospheric circulation.

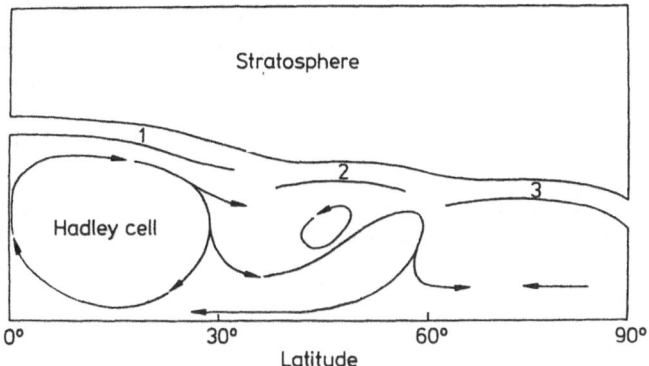

Fig. 1.3. Meridional circulation in the troposphere. Tropopause: *1*- equatorial, *2*- middle latitudional, *3*- polar.

In the upper layers of the troposphere and in the stratosphere, jet flows are generated in which the maximum wind velocities attain 100 to $150 \, ms^{-1}$. The width of these currents is several hundred kilometers and they transport great masses of air. In the winter time, west winds blow in the stratosphere, whereas in summer east winds predominate in the layers above 25 km.

The formation of various currents in the atmosphere leads to the mixing of large masses of air and the transport of the compounds emitted by various sources on the Earth's surface over considerable distances. The processes of turbulent diffusion also play a large role in the dissipation of air in the overground layer. The main reasons for the generation of turbulent, chaotic air motion with the formation of numerous vortices is the inhomogeneity of heat transfer from the Earth's surface to the atmosphere and the existence of air currents moving at different velocities. High intensity of turbulent mixing results in fast vertical transport of heat and chemical compounds.

The intensity of turbulent air mixing and hence the value of turbulent diffusion coefficient K_{dif} depends on a number of meteorological factors, mainly on the wind velocity and the thermal state of the atmosphere. For example, if the thermal stratification of air is stable, which prevents the formation of convection currents, turbulent diffusion proceeds very slowly, and when convection is considerable, it increases. The effect of thermal conditions on the processes of atmospheric diffusion can be followed from the changes in the value of K_{dif} with altitude: throughout the troposphere at a normal value of the temperature gradient ($6 \, °C \, km^{-1}$) the turbulent diffusion coefficient is approximately $10^5 \, cm^2 \, s^{-1}$, whereas in the middle layers of the stratosphere it decreases to $5 \times 10^3 \, cm^2 \, s^{-1}$. This decrease is due to the temperature inversion, an increase in temperature with altitude, preventing the formation of convective currents. This inversion beginning at a level of 20 to 30 km results from the absorption of the hard component of ultraviolet solar radiation by ozone (Fig. 1.1).

Temperature inversions are observed not only in the upper layers of the atmosphere but also appear periodically at the Earth's surface. The reasons for their formation are different. Inversions are often caused by the flow of warm air over the lower cold layers. Precipitation inversions are generated when large masses of air in anticyclones are compressed and heated. Inversions near the ground with a thickness of up to several hundred meters usually appear on windless nights when the Earth's surface and the adjoining air layer are greatly cooled.

The development of temperature inversions preventing vertical air motion and dispersion of impurities can result in critical situations in the urban atmosphere.

1.3   Evolution of Chemical Composition of the Earth's Atmosphere

The Earth's atmosphere, as we know it at present, has gone through a long evolutionary path, many details of which have not yet been elucidated.

According to modern concepts based on the determination of the content of lead isotopes in the most ancient rocks containing uranium, the Earth was formed about 4.6×10^9 y ago from the protoplanetary gas-dust cloud dispersed in perisolar space.

In the process of gravitational compression of the gas-dust cloud, heat energy was evolved. Radioactive decay of short-lived isotopes, such as ^{26}Al and long-lived isotopes of uranium, thorium, potassium and rubidium, also led to gradual heating of the planet's interior. The tidal deformations caused by the gravitation of the Moon formed simultaneously with the Earth also served as an additional source of heat. The evolution of thermal energy led to at least partial melting of the planet's material and to the evolution of gaseous substances both those existing initially in the form of ice and those formed as a result of a number of chemical processes.

Hence, a dense atmosphere around the Earth was formed from the vapor and gases evolved as a result of the degassing of the interior of the Earth. It is supposed that the gases erupted from volcanoes during the first 500 million of years of the existence of our planet consisted mainly of hydrogen, water vapor, methane and carbon oxides with an admixture of some sulphur compounds. Four thousand million years ago the condensation of water vapor led to the formation of the hydrosphere. Already at this stage considerable changes in atmospheric composition had evidently occurred. Apart from the continued degassing of the mantle, the most important factors of the early evolution of the atmosphere were such processes as the dissolution of carbon dioxide in reservoirs, the dissipation of light gases, hydrogen and helium, into the space and, as a result, an increase in the relative nitrogen content.

The most important feature of the atmosphere of that period was the absence of free oxygen. One can say with certainty that oxygen was not emitted into the atmosphere from the red-hot bowels of the planet just as it is also absent among the gases erupted from contemporary volcanoes. The formation of some amounts of oxygen could occur as a result of the photodissociation of water under the influence of ultra-violet solar radiation. However, it could not accumulate in the atmosphere containing large amounts of reducing gases (H_2, CH_4, H_2S, SO_2, etc.) and coming into contact with the reduced rocks of the earth crust.

Hence, an anaerobic reducing atmosphere quite unlike the contemporary atmosphere probably existed in the early stages of evolution. When it was finally transormed into the oxidative and aerobic atmosphere, the vital activity of photosynthesizing organisims was the factor responsible for this transition. The appearance of life on Earth had to be preceded by the accumulation of large amounts of organic substances. Many experiments have shown that under anaerobic conditions the abiotic synthesis of numerous simple and complex organic compounds is possible. The molecules of methane, carbon oxide, hydrogen, water and ammonia could be the initial molecules for this process. The main sources of energy were ultraviolet solar radiation (for which the reducing atmosphere containing no oxygen and ozone was transparent) and atmospheric electric discharges [2, 3].

Model experiments on the effect of ultraviolet rays, electric discharges and ionizing radiation upon the gas mixtures containing the above components of the hypothetical ancient atmosphere in different ratios have invariably led to the formation of hydrocarbons, aldehydes and ketones, C_1–C_3 carboxylic acids and amino acids. Considerable amounts of volatile organic compounds and amino acids are also formed during short-term contact of the same mixtures with the material of volcanic lava heated to 900–1000 °C. On the basis of this fact, many researchers supposed that large-scale abiogenic synthesis of organic substances from simple gases could proceed in the ash-gas volcanic plumes [4a].

In the prebiological period of Earth's existence, the abiotic synthesis evidently led to the formation of large amounts of organic matter. On the basis of data on the values of the quantum yield of the photochemical synthesis of amino and oxyacids from simple gases. K. Sagan [4b] has calculated that during 1×10^9 years so much organic matter was formed that if it were dissolved in the contemporary ocean, it would produce a 1% solution.

Neglecting the problem of the true concentration of organic compounds in the atmosphere and in the water of the ancient ocean, we will indicate that this concentration was sufficient for the appearance and maintenance of primitive forms of life in the early stages of evolution. This evidently happened 3.8×10^9y ago or even earlier: this is the age of rocks containing organic microstructures which may be the remains of primitive microorganisms [2]. In the sedimentary rocks of the Swaziland system (South Africa) the age of which is 3.4×10^9 y, the remains of already formed microorganisms were discovered.

The living organisms that appeared in the water of the ancient ocean became the most important factor for further evolution of the atmosphere. The main result of their activity was the accumulation of large amounts of free oxygen accompanied by the extraction of carbon dioxide. The problem of the time when the transition from the anaerobic reducing atmosphere to the aerobic and oxidative atmosphere occurred is solved on the basis of the study of fossil remains of ancient organisms and changes in the processes of formation of sedimentary rocks. Many species of multicellular organisms some of which exist at present were discovered in the deposits of the upper Pre-cambrian the age of which is 0.6×10^9 y. This fact indicates that at that time the oxygen content of the atmosphere was high: it was at least 20% of the present value. In still earlier sedimentary rocks the remains of aerobic eukariotic organisms were found. Their cells contain a nucleus as well as mitochondria and chloroplasts separated by a system of membranes.

The information about the oxidative state of the atmosphere in the earlier geological periods is provided by some ferrous formations (the so-called red beds) formed on land as a result of metamorphic changes and the oxidation of iron-containing rocks from the effect of atmospheric air and surface water. The age of the earliest of these formations is 2×10^9 y. This suggests that the transition to the oxidative atmosphere took place not latter than 2×10^9 y. However, according to the opinion of many researchers, this transition is separated from the beginning of the activity of photosynthesizing organisms by a considerable period of time.

In many lime deposits, stromatoliths have been discovered: they are carbonate structures formed as a result of the vital activity of colonies of lower plants, mainly the blue-green algae. The age of some of them is 3×10^9 y. The reason for the time break between the beginning of photosynthesis accompanied by the assimilation of carbon dioxide with the emission of oxygen and the transition of the atmosphere into the oxidative state is evidently the fact that the water of the ancient ocean contained large amounts of Fe^{+2} ions. Soluble compounds of bivalent iron formed a part of the minerals of the Earth's crust and were involved in migration processes as a result of weathering and erosion. Even the fact of transport of Fe^{+2} compounds in dissolved form indicates that an anaerobic environment existed. On the other hand, the presence of a source of photosynthetic oxygen is a condition of generation of ore formations consisting of insoluble compounds of trivalent iron. Ferrous formations have been dis-covered in the most ancient of the investigated rocks from Isua (Greenland) the age of which is 3.8×10^9 y. This is probably the time of the beginning of the activity of autotrophic photosynthesizing organisms. They replaced the earlier forms which had obtained energy by the assimilation of organic matter formed during abiotic synthesis. They also replaced photosynthesizing bacteria that had used the solar energy for the oxidation of reducing gases of volcanic origin.

Hence, the history of the evolution of the Earth's atmosphere and hydro-sphere at this stage which continued for 2×10^9 y is a process of slow titration by oxygen of the reduced components of the environment. The transition to the aerobic atmosphere began when the rate of the oxygen evolution exceeded that of the uptake of Fe^{+2} by the ocean [4c].

Different concepts exists on the dynamics of oxygen accumulation in the Earth's atmosphere. Some authors believe that after the transition to the aerobic atmosphere, rapid (on the scale of geological time) accumulation of oxygen

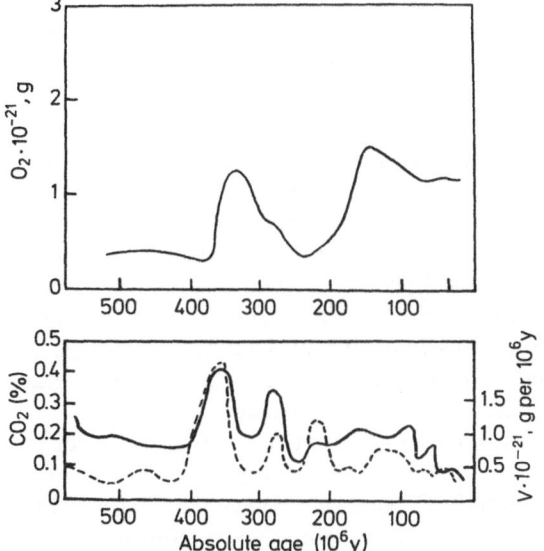

Fig. 1.4. Change in oxygen and CO_2 content in the atmosphere in the Phanerozoic. *Broken line* in the lower diagram shows a change in the velocity (v) of accumulation of volcanogenic rocks (Budyko M.I., 1980)

approximately up to the present level took place. However, the most widespread opinion is that the amount of oxygen in the atmosphere increased gradually and that during the last 0.5×10^9 y considerable variations in its mass have occurred. Figure 1.4 shows the possible course of changes in the content of oxygen and carbon dioxide beginning from the Cambrian period. The calculations of the amounts of oxygen and CO_2 were based on the data concerning the rates of formation on the continents of sedimentary rocks containing organic carbon and carbonates [5].

At present, the atmosphere contains 1.2×10^{21} g of oxygen formed almost exclusively by photosynthesis. This process may be expressed in a simplified form by the equation: $CO_2 + H_2O = CH_2O + O_2$. If it is taken into account that photosynthesis yields 2.3×10^{17} g of dry organic substance annually [16], the amount of oxygen emitted by autotrophic organisms can be easily calculated. It is approximately 2.5×10^{17} g. Almost all this quantity is used in the processes of respiration and fermentation of organic matter by heterotrophic non-photosynthesizing organisms, mainly by microbes. Thus, the carbon dioxide utilized by the plants in photosynthesis returns to the atmosphere. The amount of oxygen which is not consumed by heterotrophic organisms is determined by the mass of organic substance introduced into the sedimentary rocks and excluded from the global circulation of carbon compounds for a long time. The rate of accumulation of organic carbon in the sediments has been estimated to range from 14×10^{12} to 30×10^{12} g y^{-1}.

Another source of oxygen in the atmosphere, the dissociation of water molecules, has a negligible effect on its overall balance. It is considered that about 2×10^{12} g of oxygen per year is formed in the course of photodissociation. Hence, the production and consumption of oxygen occur in a virtually closed cycle of synthesis and microbiological destruction of organic substance in the biosphere. This fact suggests that the composition of the atmosphere is to a considerable extent controlled by the biota: the living matter of the planet.

If the atmospheric composition remains relatively constant for many millions of years, this constancy is a result of the activity of the biota. In this connection, Margulis L. and Lovelock J. E. have suggested that the composition of the Earth's atmosphere is controlled by biological processes occurring on its surface in the direction of the optimization of the conditions of development of the biosphere as a whole (the so-called "Gaia" hypothesis). The idea of these authors about the homeostasis existing on the Earth: a self-regulating system capable of maintaining the environmental conditions favorable for the existence of life, is not new and is a paraphrase of the well-known V. I. Vernadsky's thesis: "Life creates in its environment the conditions favorable for its existence".

Budyko M. I. et al. [5] think that the hypothesis "Gaia" advanced in ref. [6] is not sufficiently justified. In their opinion, the changes in the chemical composition of the atmosphere essentially depend on two external factors which are not related to activity of living organisms: the evolution of the Sun leading to a gradual increase in the radiation flux and the evolution of the Earth manifested in a gradual decrease in the process of degassing of the upper mantle. Hence, the authors of ref. [5] consider the existence of the biosphere to be a result of fortuitous agreement between the processes of evolution of the Sun and the Earth. There is naturally no sense in refuting the role of evolution of the Sun and the inner layers of the Earth in the change of the composition of the atmosphere, but it is hardly possible to explain the stability of the biosphere on the basis of these factore alone. On the other hand, the hypothesis of the existence of the global homeostasis requires more detailed development, primarily the search for the mechanisms of feedbacks in the biota-atmosphere system responsible for the maintenance of the state of the quasistationary equilibrium when the external conditions vary. The establishment of these mechanisms would also help to determine the limits of the stability of the biosphere. This is particularly important at present when human activities have become the most important nature-forming factor.

The scientific and technological revolution of the second half of the 20th century, characterized by extremely high growth rates of production has resulted in rapid changes in many essentially important parameters of the environment. Moreover, these changes are moving in a direction unfavorable for humanity. Two major factors affecting the evolution of the chemical composition of the atmosphere are the overall increase in the scale of the utilization of natural resources and the appearance of new compounds intruding into the well-balanced natural mechanisms of self-purification of the atmosphere and, possibly, perturbing these mechanisms.

The possiblility of changes in the thermal conditions of the atmosphere and the Earth's surface and, finally, of the planet's climate as a result of human activities is causing great anxiety. It has already been mentioned in Sect. 1.2 that the thermal radiation of the Earth's surface is almost completely absorbed in the atmosphere by water vapor, and only a small part of this radiation is dissipated and leaves the atmosphere through a "window" in the range of 16.7 to 7.6 μm. However, this window becomes narrower as a result of absorption of IR radiation by carbon dioxide molecules at wavelengths ranging from 16.7 to 13.7 μm. The significance of CO_2 in the generation of the greenhouse effect is universally recognized. Therefore, during the past decades its content in air has been continuously checked in various regions of the planet. The most complete data have been obtained in the background laboratory on the Mauna Loa mountain on the Hawaiian Islands. These investigations have established that the average carbon dioxide concentration in the atmosphere is continuously increasing (by approximately $0.5 \pm 0.2\%\,y^{-1}$) and undergoes regular seasonal fluctuations. The amplitude of these fluctuations is about 2% of the average value and is also gradually increasing. Several thousand years ago the carbon dioxide content in the atmosphere was much lower. According to the data on the composition of air occluded in the ice cores of the Antarctic continent and Greenland, in the pre-industrial period the concentrations of CO_2 were 260–280 ppm [7].

The increase in CO_2 content observed at present probably began in the middle of the last century and is due to several factors. First, it results from the ever increasing use of fossil fuel: coal, natural gas and oil. Second, according to the opinion of some specialists, the concentration of carbon dioxide should increase because the mass of land vegetation decreases. In the developing countries of South America and Africa the most highly productive forests are being cut down and burned and the area destroyed per annum has been calculated as 1.2 $\times 10^5\,km^2$. The speed of forests clearing is continuously increasing, and it may be expected that if these rates of forest clearing are maintained, by the end of the century the forest area in these countries will have decrease by 20 to 25%. An indirect consequence of involving new territories in industrial and agricultural activities is an increase in the rate of circulation of organic material which has previously been retained in the soil for a long time. At present about 14×10^3 Tg* of CO_2 per year are emitted into the atmosphere as a result of burning fossil fuel. The destruction of forests and land ploughing should lead to additional emission of about 1×10^3 Tg of carbon dioxide. According to ref. [8], the emission of CO_2 caused by the change in the continental biomass in the period from 1860 to 1981 was $(68 \pm 14) \times 10^3$ Tg based on carbon (during the same period of time the total emission of carbon contained in CO_2 was estimated to be 167.6×10^3 Tg).

Comparison of data on emission with the experimentally found change in CO_2 content in the atmosphere shows that it increased by less than 50% of the amount that should have been expected if all carbon dioxide of anthropogenic origin had been accumulated in air. Mechanisms which partly compensate excessive CO_2 emission evidently exist in nature. One of them might be an increase in photosynthesis leading to the assimilation of larger amounts of

---

*1 Tg $= 10^{12}$ g $= 1$ million tons.

carbon dioxide. However, the buffering capacity of the biosphere is not infinite, and a further increase in pressure may result in a number of irreversible changes in the environment. In particular, it is predicted that a 60% increase in the present CO_2 content of the atmosphere may cause an increase in the temperature of the Earth's surface by 1.2 to 2.0 K.

Strong absorption of infrared radiation of the Earth's surface, which is not sorbed by the water vapor molecules is also characteristic of many other minor components of atmospheric air. They include such substances as N_2O, SO_2, NH_3, O_3 and also methane and other organic substances. The content of the latter is very low and usually does not exceed $1 \times 10^{-6}\%$ by volume. However, great attention is being directed to them in connection with the problems of health protection and also because they take part in many meteorological phenomena. Entering the atmosphere, organic substances usually become involved in a long chain of transformations accompanied by energy absorption and release. Hence, they are thermodynamically active air components. Some organic compounds are of anthropogenic origin, and others are derived from natural sources but the scale of their emission is often profoundly affected by human activities, in particular, the intensity of agricultural and industrial production.

The calculations carried out in recent years show that even a relatively slight increase in the content of methane and its nearest homologues, chlorofluoromethanes and some other halogen-containing compounds in the atmosphere can lead to an additional heating of the Earth's surface. Table 1.1 gives the main results of the calculations carried out by Wang W.C. et al. and concerning the

Table 1.1. Greenhouse effect occurring with increasing concentration of some minor atmospheric components

Component	Centre of absorption band, μm	Increase in concentration as compared with current values by n times	Increase in the average temperature of the Earth's surface *, K
CH_4	7.66	2	0.20–0.28
C_2H_4	10.5	2	0.01
CF_2Cl_2	9.13, 8.68, 10.93	20	0.36–0.54
$CFCl_3$	9.22, 11.82	20	
CCl_4	12.99	2	0.01–0.02
CH_3Cl	13.66, 9.85, 7.14	2	
N_2O	7.78, 17.0. 4.5	2	0.44–0.68
CO_2	15.0	1.25	0.53–0.79
SO_2	8.69, 7.35	2	0.02–0.03
NH_3	10.53	2	0.09–0.12

*The lowest value was obtained at a fixed altitude, and the highest value was determined at a fixed temperature of the upper cloud boundary

change in the values of the greenhouse effect occurring with increasing concentration of some minor components of the Earth's atmosphere [9]. According to these calculations, the doubling of methane concentration should lead to an increase in the temprature of the Earth's surface by 0.2 to 0.3 K. It should to noted that the climatologists regard a change in the average global temperature of the surface to be considerable if it exceeds 0.1 K and is maintained for a prolonged period of time. Its increase by 1 K predicted for the case of simultaneous doubling of the contents of methane, nitrogen monoxide and ammonia should lead to a drastic change in climate.

Chlorofluoromethanes, CCl_2F_2 and CCl_3F, can provide a considerable contribution to the greenhouse effect. At present the background concentration of these compounds is not high (0.2–0.3 ppbv). However, its twenty-fold increase as a result of human activities should lead to an increase in the surface temperature by 0.4 to 0.5 K. Measurements undertaken latter [10] have shown that the intensities of vibration bands of CCl_2F_2 and CCl_3F in the "transparent window" were higher by 17 and 5%, respectively, than those used by Wang W.C. et al. in the calculation of the greenhouse effect. This fact implies that chlorofluorohydrocarbons can have an even more marked effect on the thermal conditions in the atmosphere and on the Earth's surface.

Dickinson R.E. and Cicerone R.J. [11] have analyzed the dynamics of the accumulation of CO_2, methane, and CFMs in the atmosphere and have come to the conclusion that by the year 2050 the probable heating due to the greenhouse effect will be 1 to 5 K or even higher.

The evaluation of the greenhouse effect caused by different components and the construction of quantitative models for the prediction of climatic variations is a difficult problem because many physical and physicochemical processes should be taken into account. These calculations require the accumulation of many initial data about the sources of organic components, their concentration in various layers of the atmosphere and the mechanisms of chemical transformations. Only when this information becomes available, may we hope to predict reliably the short-term and remote consequences of the accelerated evolution of the Earth's atmosphere.

2 Time-Dimensional Distribution of Organic Components of the Atmosphere

The rapid development of modern effective methods of investigation in the last decades has led to a qualitatively new level of knowledge of the Earth's atmosphere. At present it is possible to study an extremely complex mixture of organic compounds present in negligible amounts. It has recently been established that the air near the ground contains many hundreds of organic substances with the number of carbon atoms ranging from one to forty. The accumulation of information about their composition is of fundamental importance and should increase the comprehension of the role played by small chemical components in the evolution of the atmosphere and the biosphere as a whole.

This chapter sums up the data on the qualitative composition and the concentration of approximately five hundred organic substances belonging to different classes: hydrocarbons, their derivatives containing oxygen, nitrogen, sulfur and halogens as well as organometallic compounds. Information is provided both about the background concentrations and about the content of components in the urban atmosphere in the form of vapor and contained in aerosol particles. Particular attention is directed to the compounds playing an important role in the greenhouse effect and in the atmospheric photochemical processes (lower hydrocarbons, some of their derivatives and terpenes) and components particularly harmful to human health (lead- and mercury-containing substances, polynuclear aromatic hydrocarbons, etc.).

2.1 Hydrocarbons

2.1.1 Methane

Methane is one of the major organic components of the Earth's atmosphere. Its emission as a part of marsh gas, fire-damp and natural fuel gas had been known for a long time but the presence of methane throughout the troposphere was established by Migeotte [17b] only in 1948. In the past decades a great number of investigations were concerned with the determination of methane content in air and of its main sources. This interest is due to the recognition of a considerable significance of this constant component in various atmospheric processes. For example, it has been established that it participates in ozone formation in the troposphere as a result of photochemical reactions. Some researchers regard the oxidation of methane in the stratosphere as one of the main sources of carbon oxide, water vapor and very important intermediate particles: hydroperoxide radicals — HO_2.

At present the data (partly given in Tables 2.1 and 2.2) on the methane content in the overground air layer over the continents of both hemispheres and in the atmosphere above seas and oceans have been obtained. As can be seen from Table 2.1, methane concentrations in the air of the Northern hemisphere vary over a relatively wide range 1.4 to 2.3 ppm, however, more often they range from

Table 2.1. Average methane concentrations in the lower troposphere over the continents

Location of observation	Latitude	Date of observation	Conc., ppm	References
Northern Hemisphere				
USA:				
Alaska, Point Barrow	71°	August–September 1967	1.59	12
		October–December 1980	1.66	13
Nebraska, 1200 m above sea level	42°	1962–1967	1.69 ± 0.03	14
Colorado, 2000 m above sea level	40°	1969	1.62	15
New Mexico, 2860 m above sea level	33°	April 1984	1.59	16
Florida	28°	January 1978	1.63	17a
USSR:				
region of city Vorkuta	67°	March 1976	1.8 ± 0.2	18
Yakutsk Autonomous SSR	58.5°	June–July 1975	1.70	19
Novosibirsk distr.	54.7°	May–July 1974	1.70	19
Amur distr.	53.5°	July 1975	1.70	19
Turkmenian SSR	38°	March 1978	2.3 ± 0.2	18
France	45°	July, September 1977	1.68	20
Saudi Arabia	26°	Summer of 1979	1.59	21
India, Ganges valley	25°	July 1979	1.63	21
Southern Hemisphere				
Antarctica, South Pole	90°	January 1979	1.48	22
		January 1980	1.50	
		January 1981	1.52	
		January 1983	1.55	
		January 1984	1.57	

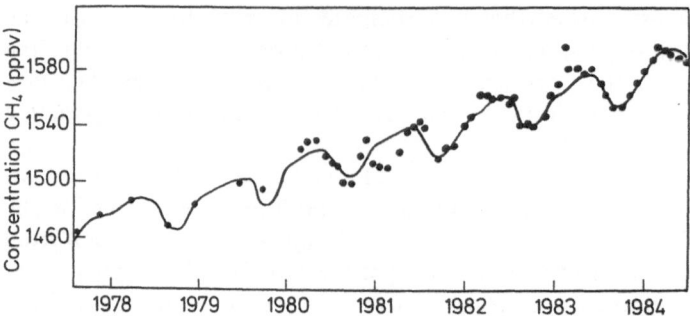

Fig. 2.1. Monthly mean CH_4 concentrations observed at Cape Grim, Tasmania [26]

1.6 to 1.7 ppm. The highest CH_4 content sometimes attaining 5 ppm is observed in the air near the ground over territories which contain large deposits of oil, natural gas or coal and over seismically active regions [18]. The lowest concentrations have been observed in the atmosphere of high mountain regions. The mean background methane concentration in dry air measured in the middle of the 1980s was 1.625 ppm [23].

The determination of background concentrations in the overground air has permitted the establishment of slight seasonal variations: the amount of methane decreased in winter and summer and two maxima were observed in spring and late autumn [13, 23–25]. In recent years, data on the seasonal fluctuations of methane concentrations were obtained in both the Southern and the Northern hemispheres. According to the results of measurements at Point Barrow (71 °N) undertaken in 1980–1982, the amplitude of variations in CH_4 concentrations between the summer and autumn months was on the average 34 ppb, whereas at a background station in Oregon (45 °N) in 1979–1982 it was 27 ppb and in the South of Tasmania (Cape Grim, 41 °S) it was 28 ppb [13, 26].

Considerable variations in methane concentrations in the air layer near the Earth's surface over land indicate that its content is profoundly affected by a number of local factors. Since it is necessary to exclude this effect from the results of analyses, considerable research has been carried out on the determination of background methane concentrations in the air over the seas and oceans, the waters of which, beyond the continental shelf, are a very weak source of hydrocarbons.

Table 2.2 gives the results of these investigations carried out in both hemispheres. Attention is drawn to the existence of a concentration gradient in the troposphere between both hemispheres: the sea air of mid-latitudes of the Northern hemisphere is characterized by higher (by 8 to 10%) methane content than that in the same latitudes of the Southern hemisphere. The most pronounced difference in concentration is observed in the region of the intratropical convergence zone (about 5 °N) separating the trade winds of both hemispheres and characterized by intensive convection. As will be clear from the next chapter, the existence of this gradient is due to the fact that the amount of methane

Table 2.2. Average methane concentrations in the lower troposphere over seas and oceans

Location of observation	Latitude	Date of observation	Conc., ppm	Refer- ences
Northern Hemisphere				
Norwegian and Green- land Seas	80–55°	August 1971	1.37 ± 0.05	27
Northern Atlantic	54–50°	June 1971	1.57–1.73	28
Eastern Atlantic	58–12°	June 1978	1.72 ± 0.13	29
Central Atlantic	40–14°	October– November 1980	1.68–1.70	30
Caribbean Sea	25–20°	June–July 1973	1.85	
Pacific Ocean:	35–10°	November 1972	1.74	28
	37–20°	April 1978	1.72 ± 0.01	31
	5°	November– December 1972	1.60	28
Arabian Sea	20°	Summer 1979	1.54	21
Southern Hemisphere				
Atlantic Ocean	0–32°	October– November 1980	1.60	30
	19–51°	May 1978	1.55 ± 0.13	29
Pacific Ocean	10–57°	May 1978	1.62 ± 0.01	31
	5–75°	November– December 1972	1.56	28
New-Zealand, Pacific shore	43.5	January 1972	1.41–1.44	32
Cape Grim, Tasmania	41°	October 1978	1.48	26
		October 1980	1.52	26
		October 1982	1.56	26
		October 1984	1.59	26
Antarctica, Mawson station	67°	October 1981	1.54	26
		October 1982	1.55	26
		October 1984	1.59	26
Antarctica, McMurdo station, Ross Sea	79°	December 1971	1.34–1.35	32

entering the atmosphere is greater in the Northern than in the Southern hemisphere.

In order to study the vertical distribution of methane concentrations, samples of air collected at different altitudes on board planes or with the aid of sondes and rockets have been analyzed. The first investigations carried out over the continental regions of North America and of the Pacific ocean have shown that the concentration of CH_4 is approximately constant throughout the troposphere [14, 15]. The regular distribution of methane up to an altitude of 8 to 11 km over middle and low latitudes of the Northern hemisphere were also reported in refs.

[13, 17, 21]. However, Seiler W. et al. [33] discovered a small vertical gradient over the territory of German Federal Republic. According to the data of analyses carried out in late autumn of 1976, the concentration of CH_4 decreased from 1.72 ppm at an altitude of 2 km to about 1.60 ppm at an altitude of 9 km.

The study of methane content in the lower troposphere carried out by Soviet researchers [18] has shown that it is of a distinctly seasonal type: in winter time almost equal methane distribution was observed up to an altitude of 5 km, whereas in summer a considerable concentration gradient was found in the same air layers (Fig. 2.2). The concentration of methane and its distribution in the air layer near the ground probably depend on the relationships between the source strength and the intensity of processes of atmospheric diffusion and photochemical oxidation. Hence, it may also be expected that the vertical gradient is different in different latitudes. Indeed, the authors of ref. [26] have established that at an altitude of 7 to 10 km over Southern Australia the methane content in the summer period (December) is approximately 16 ppb higher than at the Earth's surface. However, they have not observed seasonal fluctuations of CH_4 content in the upper troposphere which are very characteristic of the air near the ground. The authors of ref. [26] explain the observed increase in the methane content in the upper troposphere by the fact that CH_4 flows from the Northern to the Southern hemisphere at high altitudes and also by more intense photochemical oxidation in the lower layers of the atmosphere.

In the stratosphere, the methane content decreases further. Ehhalt D. et al. [14, 15, 28] have established that the greatest concentration gradient is at the tropopause and that the concentration was 1.58 ppm at the upper boundary of the troposphere (14.5 km at a latitude of 30 °N), whereas at an altitude of 19 km it was only 1.2 ppm. In the upper layers of the stratosphere at an altitude of 50 km it was only 0.3 ppm. The same methane content in the lower stratosphere has been found by the authors of ref. [34] but they have reported that it decreased to 0.6 to 0.3 ppm even at altitudes of 31 to 35 km. The data reported in ref. [33] also indicate the existence of a sharp concentration gradient. According to these data, the concentration decreases more than twice within the tropopause.

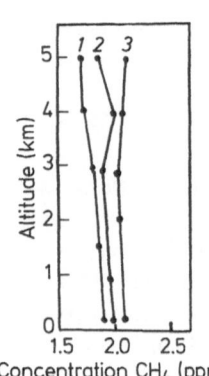 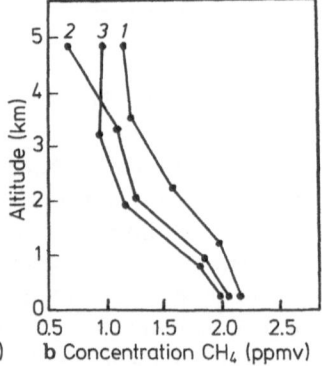

Fig. 2.2a, b. Vertical distribution profiles of methane [18]. a Winter (1976) profiles: 1- Kirov district (58°N), 2- Vorkuta district (67°N), 3- Penza district (53°N). b Concentration profiles in the spring of 1975 in Vorkuta district: 1- April 26, 2- April 27, 3- April 28

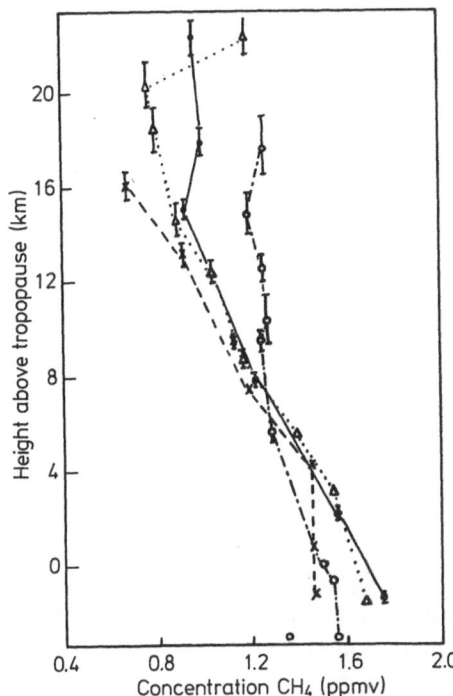

Fig. 2.3. Methane mixing ratio vs height above the tropopause. Tropopause heights: Antarctica (78°S) 8.5 km (●), Wyoming (41°N) Feb. 14, 10.4 km (△), Brazil (4°S) 17.1 km (○)

Bush Y.A. et al. [35] have observed latitudinal variations in vertical distribution of the methane content at altitudes of 6 to 35 km over the territory of the USA (41 °N), over Brazil (4 °S) and over the Antarctic (78 °S). These measurements have shown that over Brazil methane concentrations in the stratosphere decrease much more slowly than over the USA and the Antarctic. This difference is due to the existence of strong upward currents in tropical latitudes (Fig. 2.3). Schmidt M. et al. [36] reported data on the meridional distribution of methane concentrations obtained from the results of analyses of air samples collected during flights at altitudes of 8 to 11 km from Spitzbergen to Monrovia (Liberia) (6 °N). Particular attention has been directed to measurements in the vicinity of the tropopause. The authors detected a marked air exchange between the troposphere and the stratosphere, in particular near the breaks in the tropopause in polar and subtropical latitudes. This fact shows that the arrival in the stratosphere of minor gas components emitted at the Earth's surface (and, on the other hand, the intrusion of the stratospheric ozone into the troposphere) is due, not only to the convective part of the Hadley circulation in the tropics, but also to the exchange of "parcels" of air trapped by the folds of the tropopause over the subtropics and polar regions (Fig. 2.4).

The problem of the trends of changes in background methane concentrations in the atmosphere is very important. The figures given in Tables 2.1 and 2.2, in particular those concerning the air of the southern hemisphere, clearly show that

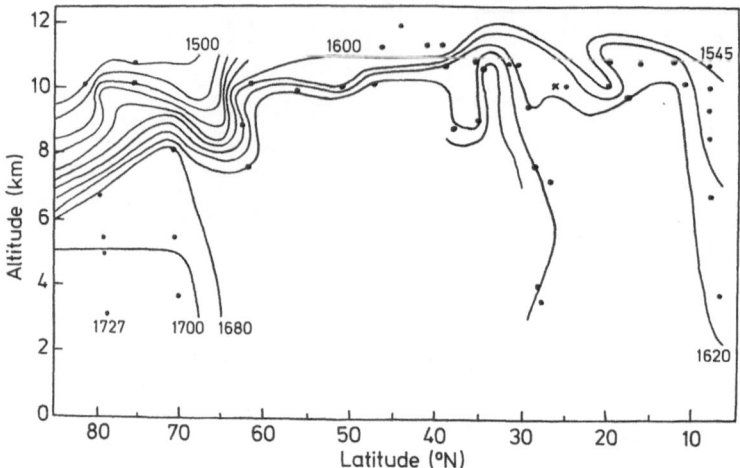

Fig. 2.4. Meridional distribution of the methane mixing ratio during flights in May 1981 [36].

these concentrations increase continuously. At present the result of continuous gas-chromatographic measurements of methane for 10 years (1975–1985) are available [22, 37]. Moreover, previously unpublished data on the gas-chromatographic and spectrometric determinations of methane in the years preceding the beginning of its systematic monitoring have appeared in the literature. Thus, the methane content in the air of a rural district in California, USA in 1968 (1.38 ppm) has been reported [38]. Methane concentration was determined from the IR solar spectra recorded in April 1951 at the Sphinx observatory in the Swiss Alps (latitude 46.5 °N, elevation 3,578 m) and repeated there in February 1981 [39]. According to the data of this paper, the methane concentration in the troposphere of the Northern hemisphere increased during the above period from 1.14 ± 0.08 to 1.58 ± 0.09 ppm. Hence, the rate of increase was on the average $1.1 \pm 0.2\% \, y^{-1}$. This rate is slightly lower than that obtained as a result of a continuous automatic control for 36 months from January 1979 to January 1982. During this control, 38,000 air samples in the West of the USA (Cape Meares, Oregon) were analyzed and the rate was found to be $1.9 \pm 0.5\% y^{-1}$. In the southern hemisphere, the methane concentration in 1979–1984 increased approximately linearly at an average rate of $1.2 \pm 0.1\% \, y^{-1}$ [22].

The investigations of 80 ice samples obtained from different depths of glaciers in Greenland and the Antarctic (the most ancient samples were 3000 years old and the youngest were about 100 years old) carried out recently by Rasmussen R.A. and Khalil M.A.K. [40] have shown that the methane content in the occluded air bubbles is much lower than the present content. The long-term trend of methane reconstructed according to the data of these analyses looks as follows. For several thousand years, the concentration of CH_4 was approximately constant and was 700 ± 30 ppb, i.e. 45% of the contemporary level. Its increase began approximately 350 years ago and ranged from 0.1 to $0.3\% y^{-1}$ up

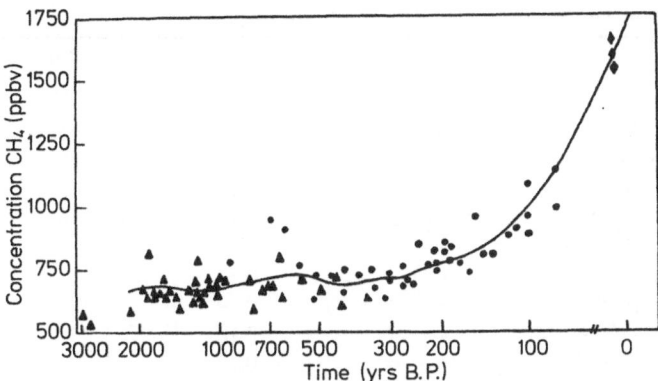

Fig. 2.5. Concentration of CH_4 in old and ancient atmospheres (100–3000 B.P.). The *circles* are data from Greenland polar ice cores and *triangles* are data from South Pole ice cores. The present concentration is shown by *diamonds* [40]

to 1800–1840. From the end of the first half of the 19th century this increase was accelerated several times. As can be seen from Fig. 2.5, for the past 100 years it is of the exponential character, which implies that this acceleration should lead to the doubling of the CH_4 content in the troposphere in the next 35–40 years.

Approximately the same features of this trend of methane have been reported by other researchers on the basis of the determination of the composition of air from Arctic ice cores [41, 42]. According to the authors of ref. [41], for 26,000 y up to the end of the 16th century, methane concentration was 0.7 ppm and then began to increase at a rate of 0.114 ppm per century attaining 1.25 ppm in 1915. From this moment the rate of increase in concentration increased by a factor of 20.

In conclusion, it should be noted that methane content in the air of large cities sometimes markedly exceeds the background concentration. According to a communication from Altshuller A.P. et al. (1966), the concentrations of CH_4 in the air of Lost Angeles and Cincinnati was on the average 2.6 and 2.5 ppm, respectively. More recent data have been reported in ref. [43]. They show that the methane content in the air of 16 large cities of North and South America, Europe and Asia usually ranges from 1.7 to 2.0 ppm and seldom exceeds these values.

Summing up the data reported in this section, it is possible to say that methane is characterized by relatively regular distribution throughout the troposphere, regular meridional distribution, a slight latitudinal gradient, relatively moderate seasonal variations of concentrations and a slight local increase in the background level in cities.

2.1.2 Nonmethane Hydrocarbons

The pattern of the time-dimensional distribution of methane described in the preceding section is also in many respects characteristic of some other minor

Table 2.3. Background concentration of acetylene in the atmosphere

Location of observation	Latitude	Date of observation	Conc., ppb	References
Atlantic ocean	81–55 °N	August 1979	0.08–0.46	45
	2.5 °N–4 °S	May 1979	0.07–0.2	45
Pacific ocean	75 °N–25 °N	November 1976	0.2	45
	40 °N–30°N	December 1981	0.46	46
	30°N–20°N	— " —	0.42	
	20 °N–10°N	— " —	0.37	
	10 °N–0°	— " —	0.09	
	0°–10 °S	— " —	0.16	
	20 °S–32 °S	— " —	0.13	
	25 °S–65 °S	November 1976	0.08	45
South Pole	90 °S	November 1977	0.06	47

gaseous components of the atmosphere having relatively long lifetimes. However, the Earth's atmosphere always contains hundreds of other hydrocarbons. The reactivity of all these compounds is much higher and the lifetime in the atmosphere much shorter, which affects the character of their distribution. Up till now, the most complete pattern of meridional, latitudinal and vertical distribution has probably been obtained for acetylene. Since it contains some features common for the entire group of low molecular weight volatile hydrocarbons, it will be considered in somewhat greater detail.

Acetylene is not one of the major organic components of atmospheric air but attracts great attention of specialists in the chemistry of the atmosphere because it is often used as a tracer characterizing the contribution of automobile transport to urban air pollution. Since its reactivity in photochemical reactions is relatively low, it is convenient for the study of the processes of dissipation and transport of anthropogenic impurities by air currents [44].

The background acetylene concentrations in the lower troposphere over the oceans are given in Table 2.3. It is noteworthy that a considerable concentration gradient exists: in the case of methane it does not exceed 10%, whereas the acetylene content decreases many times on passing from the Northern to the Southern hemisphere. This is the first difference between the contemporary distribution of shortlived minor gaseous components of the atmosphere and that of longlived components.

The second difference is the existence of a strong vertical concentration gradient of the former group of atmospheric components. In the 1970s and 1980s, many researches determined the vertical concentration profile for a number of volatile hydrocarbons including acetylene [16, 47–51]. According to the data of Cronn D. and Robinson E. [47], in the middle layers of the atmosphere over the Pacific ocean in the latitudinal range of 75 °N–25 °N, acetylene concentrations are close to 0.20 ppb, whereas in the 25 °S–65°S range they are on the average 0.10 ppb. Above the tropopause, the acetylene content in air decreases rapidly to

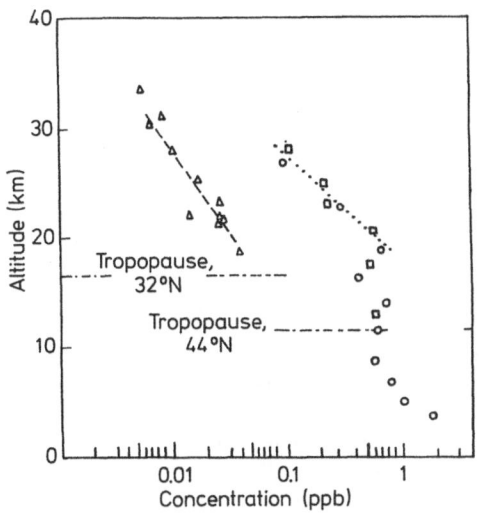

Fig. 2.6. Vertical profiles of C_2H_2 mixing ratios in the troposphere and stratosphere. *Circles*: 16 June 1979 at 44°N 2°E; *squares*: 28 June 1979 at 44°N 2°E; *triangles*: 20–21 September 1981 at 32°N 96°W. *Dotted line*: least square fit to exponential profile for data from 44°N. *Dashed line*: same as dotted line, but for data from 32°N [49]

0.01 ppb. The vertical distribution of C_2H_2 up to an altitude of 30 km over the middle latitudes of the Northern hemisphere is shown in Fig. 2.6 [49].

The meridional cross-section of concentrations of acetylene and other volatile C_2–C_5 hydrocarbons was obtained in April–May 1980 during a survey flight of STRATOZ II in the latitudinal range from 60 °N to 60 °S at an altitude from 2 to 12 km [50]. All hydrocarbons were characterized by a decrease in concentration from the South to the North and with increasing altitude. However, the decrease in concentration with altitude was not monotonous: for the nearest methane homologues and light alkenes in the middle latitudes, an increase in concentration was even observed directly under the tropopause (it was less pronounced for acetylene). This unusual gradient is due to the existence of strong upward air currents near the fronts of the intratropical convergence zone. These currents are able to transport masses of near-surface air to a considerable height.

Sometimes it has also been possible to establish the perturbations in the usual trend of concentration gradients of C_2H_2 in the lower layers of the troposphere. For example, during a flight over Wilmington, Ohio, on August 15, 1974, it was found that the concentration of acetylene dropped from 1.4 ppb in the air near the ground to 0.9 ppb at an altitude of 1000 to 1500 m, increased to 1.7 ppb in a layer of 1500 to 2000 m and then dropped again to 0.1 ppb or even lower [52]. This unusual distribution is probably observed under the conditions of temperature inversion preventing the formation of vertical turbulent currents and the dispersion of components involved in the air masses passing over the urban areas.

However, acetylene concentrations in the atmosphere of cities attain considerable values (Table 2.4) exceeding the backgroud concentrations by one or two orders of magnitude. The amount of acetylene in urban air undergoes great diurnal variations. The maximum concentrations in the cities of the USA and Western Europe are observed in the morning and evening hours. A typical

Table 2.4. Acetylene content in the atmosphere of some cities

City	Date of observation	Concentration, ppb minimum	mean	maximum	Ref.
USA:					
Los Angeles	Summer 1973:				53
	3.00–5.00	—	15	—	
	7.00–9.00	—	45	90	
	12.00–14.00	—	25	—	
New York, Manhatten,	Summer 1975:				44
at a height of 15m	5.30–8.30	—	102	—	
	12.30–15.30	—	104	—	
at a height of 308m	5.30–8.30	—	46	—	
	12.30–15.30	—	28	—	
Hamburg	Summer 1976:				54
	daytime	10	—	200	
	nighttime	5	—	20	
West Berlin	Summer 1977	0.09	—	89	56
Cities of the Rhein-Ruhr area	April 1981–March 1982	4	—	17	55
Lancaster (England)	May–June 1983	—	10.3	—	57
Sydney (Australia)	1979–1980	—	10	—	58
Bombay (India)	January–December 1982	0.6*	—	10.3*	59

*Minimum and maximum average monthly concentration in the afternoon

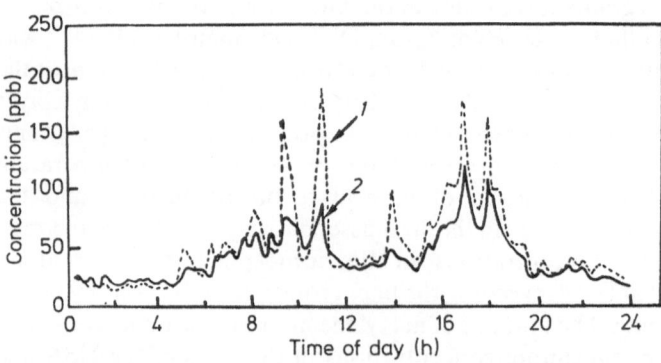

Fig. 2.7. Diurnal trend of acetylene (*1*-) and ethylene (*2*-) in urban air [54]

Table 2.5. Background concentrations (ppb) of low molecular weight hydrocarbons over oceans

Location of Observation	Latitude	Date of observation	Ethane	Ethylene	Propane	Propylene	Isobutane	Butane	Pentane	Refs.
Spitsbergen	75°55'N	March 1983	3.95 ± 0.03	0.156 ± 0.037	2.16 ± 0.28	—	—	—	—	60
Point Barrow, Alaska	71°N	Autumn 1967	0.02–0.08*	—	—	—	—	0.03–0.19	0.08–0.3	12
		March 1982	2.87	0.08	1.26	—	—	—	—	61
Atlantic Ocean	81°N–55°N	August 1979	1.0–3.3	0.01–0.51	0.13–2.5	0.01–0.20	0.02–0.47	0.02–1.3	0.02–1.3	45
	2°51'N–00°8'N	May 1979	0.8–0.9	0.1–0.2	0.0–0.2	0.09–0.1	0.01–0.04	0.0–0.1	0.01–0.06	45
	1°20'S–4°7'S	May 1979	1.0	0.4–0.5	0.1	0.0–0.1	0.02	0.0–0.1	0.02–0.03	45
Pacific Ocean	40°N–30°N	December 1982	2.37	0.12	0.80	0.05	0.21	0.51	0.42	46
	30°N–20°N	"	1.84	0.10	0.72	0.18	0.20	0.66	0.43	"
	20°N–10°N	"	0.94	0.05	0.39	0.30	0.28	0.65	0.37	"
	10°N–0°	"	0.34	0.11	0.27	0.16	0.10	0.30	0.26	"
	0°–10°S	"	0.29	0.07	0.31	0.15	0.13	0.19	0.29	"
	10°S–20°S	"	0.27	0.07	0.20	0.28	0.11	0.13	0.36	"
	20°S–32°S	"	0.23	0.08	0.11	0.07	0.06	0.14	0.17	"
	17°S–11°S	July–December 1982	0.50–0.270	0.20–4.0	0.16–0.84	0.07–0.49	—	0.08–0.18	0.01–0.16	62

*Together with ethylene

picture of the diurnal change in acetylene concentrations in urban air is shown in Fig 2.7.

The identical or even more pronounced anisotropy of distribution is also characteristic of other hydrocarbons. Table 2.5 lists data on the content of the nearest C_2–C_5 methane homologues and alkenes over the oceans and the Arctic shores. It is clear that upon passing from the Northern to the Southern hemisphere the concentrations usually decrease three or four times.

The vertical distribution of methane homologues [16, 47, 48, 50, 63–65] is characterized by the most drastic decrease in concentrations at the upper boundary of the tropopause. In the interval between the tropopause and the altitude of 30 km, the content of light alkanes drops by about three orders of magnitude.

Rudolph et al. [63] recorded a decrease in ethane concentrations from 2 ppb to 2 ppt in the interval between the tropopause and the altitude of 30 km. The concentration of propane decreased even faster: it did not exceed 10 ppt at an altitude of 18 km.

At present, the problem of the flow of acetylene and methane homolgoues through the tropopause and their distribution in the stratosphere is not an abstract question. There are reasons to suppose that although their concentrations are very low, they can affect the tropospheric chemistry of chlorine and ozone compounds to the same extent as does the more chemically inert methane.

In recent years, considerable seasonal variations in the content of volatile hydrocarbons in some background regions have been observed. They are most pronounced in high latitudes of the Northern hemisphere [60, 61, 66]. The maximum concentrations of ethane and propane in the Arctic air have been recorded in winter months (December–February), whereas the minimum concentrations have been observed in summer and early autumn (July–September). For example, ethane concentrations in Alaska vary from about 2 ppb in winter to 0.8 ppb in summer [66]. These variations are evidently caused by two main factors: the characteristic time of transport of the components by air currents from middle to high latitudes (minimum in winter) and the lifetime in the atmosphere (minimum in summer) depending on the intensity of photochemical processes [60, 61].

The sea air contains in the gas phase not only the compounds listed in Table 2.5 but also heavier hydrocarbons. Their total content is usually low and does not exceed several micrograms per cubic meter. For example, the average overall concentrations of C_2–C_7 hydrocarbons in air over the Pacific Ocean near California were $2.8\ \mu g\,m^{-3}$, whereas over the north Atlantic the sum of C_3H_8–$C_{12}H_{26}$ alkanes ranged from 4 to 6 $\mu g\,m^{-3}$. The mass of air arriving at the western shore of Ireland and having no contact with land for at least two days contained on the average $0.09\ \mu g\,m^{-3}$ of C_9H_{20}–$C_{17}H_{36}$ n-alkanes with the dispersion varying from 0.06 to 0.13 $\mu g\,m^{-3}$. Almost the same concentrations of these hydrocarbons have been observed in air coming from the Indian ocean to Tasmania [67].

Much larger amounts of volatile organic susbtances are present over the continents. The hydrocarbon background of air in rural areas is summed up from the emissions of local natural sources and pollutants emitted from urban areas. The total amount of hydrocarbons in air depends on the contribution of both sources and can vary by more than one order of magnitude over a short period of time. Thus, according to the reports of several authors, the total hydrocarbon content in the non-urban air in the USA in 1973–1977 usually ranged from 10 to 120 ppb. For example, the concentrations of approximately 40 C_2–C_{10} hydrocarbons in a swamp district Everglades in the south of Florida in May 1976 were 94ppb based on carbon at ground level and 62 ppb at an altitude of 0.6 km [68].

The most extensive information on the content of volatile organic substances in the atmosphere of non-urban areas has been obtained for the American continent. Little information is available for rural areas of Eastern Europe. The analyses carried out by the author in the autumn of 1981 and in the spring of 1982 in a rural area 150 km north of Vologda showed that the total amount of C_4–C_{10} hydrocarbons ranges from 20 to 60 ppb and is profoundly affected by meteorologic conditions. The highest concentrations were observed for the south and south-western wind, and moreover the relative content of C_6–C_8 aromatic hydrocarbons which form a part of vehicle fuel increased. The presence of oil hydrocarbons indicates that anthropogenic sources profoundly affect the chemical composition of the atmosphere even in non-industrial areas with a low population density at a considerable distance from large cities.

In recent years, attention has been increasingly focused on the investigation of the qualitative and quantitative composition of organic compounds emitted by natural sources and mainly by vegetation. The interest in biogenic components is associated with the suggestion that they participate in many important atmospheric processes. This refers primarily to such reactive compounds as C_5H_8 isoprene and $C_{10}H_{16}$ monoterpene hydrocarbons which under certain conditions can take part in photochemical reactions leading to the formation of ozone and aerosol particles (see chap. 5).

In the air under the cover of forests of various types, more than 15 monoterpene and sesquiterpene hydrocarbons have been found [62, 69–78]. They are an acyclic hydrocarbon, myrcene, and also mono- and bycyclic terpenes of the series of menthane (limonene, α- and γ-terpinenes, α- and β-phellandrenes, terpinolene), pinane (α- and β-pinenes), carane (3- and 4-carenes), isobornylane and isocamphane (α- and β-fenchenes, camphene). Sesquiterpene hydrocarbons are represented by longifolene, β-humulene and β-bisabolene [70]. The information available in the literature on the terpene content of air is very contradictory. The first attempts to carry out a quantitative determination of terpenes wcre made as early as the 1930s [79]. For this purpose a method of "wet burning" was developed. The terpenes were absorbed by passing the air under analysis through a solution of potassium bichromate in concentrated sulfuric acid. Burkser et al. [79] reported that in the air of pine forests in the Ukraine, the concentrations of terpene hydrocarbons at air temperatures of 27 to 42 °C ranged from 34 to 210 ppb and in young forests in the absence of wind attained 825 ppb. However,

Table 2.6. Content of monoterpene hydrocarbons in forest air

Location of sample collection	Forest type and age	Total terpene concentration, ppb	Ref.
USA:			
Rocky Mountains (Idaho)	High mountain pine and fir forest	0.07–1.6	80
Rocky Mountains (Colorado)	— " —	0.27–0.63	76
Adirondack Mountains (New York)	Coniferous forest with predominance of fir	0.3–8.2	73
Norway	— " —	8.8–70.7	75
USSR:			
Byelorussia	Pine plantation, 30–40 y	0.2–2.4	71
Ukraine	Pine plantation, 20–25 y	1.1–4.4	71
Vologda district	Pine forest, 250 y	0.6–5.8	71
Georgia	Mountain pine forest	8–50	81
France, Atlantic coast	Pine and fir forest	0.5–7.5	82
Japan, Tsukuba:	Pine forest	0.05–2.1	74
	Cryptomeria forest	0.2–0.5	
Brazil	Tropical forest:		62
	under the canopy	1.75–10.7	
	in an air layer	85	
	from treetop to 2 km		

the data obtained in recent years with the aid of gas chromatography show that the content of terpenes in the air of forests is much lower and does not exceed 70 ppb (Table 2.6).

According to the communication of Roland L. and Mohnen V.A. (1978), the ranges of individual concentrations of α-pinene, β-pinene and limonene in the coniferous forests of Whiteface Mountain, New York (1490 m elevation) in summer are 0.07 to 4.0, 0.54 to 3.6 and 0.21 to 15 ppb, respectively. The concentration of p-cymene genetically related to terpenes of the menthane series was 0.50 to 24 ppb.

Probably the method of analysis used in the 1930s yielded excessive results because not only terpenes but also other organic materials constantly present in the air of forests (plant pollen, spores of microorganisms, etc.) were oxidized. On the other hand, the results of the modern methods of analysis may give slightly underestimated concentration values because terpenes are partly lost in the preparatory stages. When samples are collected in plastic bags or metal containers, as is often done by American researchers, organic compounds can be partly sorbed on the walls and in various communication lines. It cannot be ruled out that terpenes possibly undergo oxidation upon storage in these containers by interacting with ozone molecules. The background concentration of ozone in the lower layers of air is 20 to 25 ppb and in some cases attains 70 to 100 ppb. When terpenes are concentrated in sorption tubes, reaction with ozone can also proceed on the sorbent surface. Finally, the observed variations in terpene content under the cover of forests can be caused not by the methods of analysis but, rather, by the differences in the meteorological conditions at the time of sample collection: the inverse relationships between terpene concentration under the forest cover and the wind velocity was mentioned as early as 1940 [79]. The latter reason seems to us the most important of them, and the effect of atmospheric diffusion on terpene content in air will be considered in greater detail in Chap. 3.

Almost all the data on the content of biogenic organic compounds have been obtained for the forests of middle latitudes of the Northern hemisphere, and almost no information is available on their concentration in the air of highly productive evergreen tropical and equatorial forests of America, Africa and Asia. We know of only two communications about the concentrations of organic substances under the cover of tropical forests in Brazilian Amazonia and they show great discrepancy. According to the data of Greenberg J.P. and Zimmerman P.R. [62], the total concentration of C_2–C_{10} hydrocarbons under the cover of forests did not exceed 30 ppb (about $83\,\mu g\,m^{-3}$). In an air layer from the treetops to 2 km it was usually larger but did not exceed 35 ppb (about $100\,\mu g\,m^{-3}$). However, the investigation of Sheesbey D.C, et al. cited in ref. [83] has shown that the amount of organic compounds during the wet season is very great and on the average exceeds $1500\,\mu g\,m^{-3}$. These concentrations are approximately 500 times higher than those over the oceans and are close to those observed in the air of large cities.

Volatile C_1–C_2 hydrocarbons make up the major fraction of the organic component of the air basin of cities. The total concentrations of nonmethane hydrocarbons in the urban atmosphere often exceed those in the non-urban areas by 2 or 3 orders of magnitude. The highest concentrations are observed in the cities located in hollows or protected by mountains from winds because prolonged windless periods and frequent temperature inversions favor the accumulation of pollutants.

Judging from the few published data, hydrocarbon concentrations in the cities of Western and Eastern Europe are of the same order of magnitude as those in the USA. For example, in three districts of West Berlin the mean concentrations of 40 C_2–C_{12} hydrocarbons in 1976 were 116, 133 and $756\,\mu g\,m^{-3}$ [56]. The total content of 13 hydrocarbons in the area of London Heathrow airport in 1977 was on the average $900\,\mu g\,m^{-3}$ [84]. A single measurement of the content of 21 C_7–C_{12} hydrocarbons in Zurich yielded the value of $1112\,\mu g m^{-3}$ [85].

Table 2.7 contains information on the C_1–C_{20} hydrocarbons usually present in the urban air in concentrations not lower than 0.01 ppb. This list is drawn up from the results of air analyses in more than 20 cities in North America, Europe, in the Asian territory of the USSR, in South Africa and Australia. Apart from the

Table 2.7. Hydrocarbons found in the urban atmosphere

Compound	Formula	Mol. wt.	Concentr., ppb	References
Methane	CH_4	16	1600–7000	43
Acetylene	C_2H_2	26	10–502	32, 54, 56, 58, 93
Ethene	C_2H_4	28	10–152	54, 56, 58, 89, 93
Ethane	C_2H_6	30	0.1–91	56, 58, 68, 86, 89
Propyne	C_3H_4	40	0.5	58, 93
Allene	C_3H_4	40	—	93
Propene	C_3H_6	42	16–70	56, 58
Cyclopropane	C_3H_6	42	—	93
Propane	C_3H_8	44	3–97	56, 58, 68, 86, 89, 93
1,3-Butadiene	C_4H_6	54	—	89, 93
1-Butene	C_4H_8	56	0.04–6.8	58, 86, 87
2-Butenes (*cis-*, *trans-*)	C_4H_8	56	1.2–6.8	58, 68, 85, 89, 90, 92, 93, 96
2-Methylpropene	C_4H_8	56	2.4–14	56, 58, 68, 87
2-Methylpropane	C_4H_{10}	58	1.5–53	56, 58, 68, 89, 93, 95
Butane	C_4H_{10}	58	5.0–276	56, 58, 89, 92–94, 96–100
Isoprene	C_5H_8	68	0–2.6	68, 92, 93
Cyclopentene	C_5H_8	68	0.3–9.9	93
2-Methyl-1-butene	C_5H_{10}	70	0.5	58, 93
2-Methyl-2-butene	C_5H_{10}	70	1.3	58, 93
2-Methyl-3-butene	C_5H_{10}	70	—	93
1-Pentene	C_5H_{10}	70	0.3–81.3	58, 91, 92, 96–99
2-Pentenes (*cis-*, *trans-*)	C_5H_{10}	70	0.3–3.8	58, 68, 93
Cyclopentane	C_5H_{10}	70	0.3–48	56, 58, 92, 93, 95, 99
2-Methylbutane	C_5H_{12}	72	0.9–105	58, 89, 91–100
Pentane	C_5H_{12}	72	5.0–185	56, 58, 87, 89, 91–100
Benzene	C_6H_6	78	0.06–205	56, 58, 84–100
1-Hexene	C_6H_{12}	84	0.08–53.4	91, 92, 99
3-Hexenes (*cis-*, *trans-*)	C_6H_{12}	84	0.1–0.3	68, 93
2-Ethyl-1-butene	C_6H_{12}	84	—	93
2,3-Dimethyl-1-butene	C_6H_{12}	84	—	93
2,3-Dimethyl-2-butene	C_6H_{12}	84	—	93
2-Methyl-2-pentene	C_6H_{12}	84	—	93
3-Methyl-2-pentene	C_6H_{12}	84	—	93
4-Methyl-2-pentene	C_6H_{12}	84	0.1–4.5	68, 93
Methylcyclopentane	C_6H_{12}	84	0.3–59	56, 58, 87, 91–93, 96–100
Cyclohexane	C_6H_{12}	84	0.8–75	56, 58, 90–95, 100
2,2-Dimethylbutane	C_6H_{14}	86	0.5–5.5	56, 58, 92, 96–99
2,3-Dimethylbutane	C_6H_{14}	86	0.5–5.5	56, 58, 87, 92, 93, 96–98
2-Methylpentane	C_6H_{14}	86	1.8–106	56, 58, 90–99
3-Methylpentane	C_6H_{14}	86	1.3–73	56, 58, 91–99
Hexane	C_6H_{14}	86	1.8–123	56, 58, 87, 90–100
Toluene	C_7H_8	92	0.1–156	56, 58, 84–100

Table 2.7. (*Continued*)

Compound	Formula	Mol. wt.	Concentr., ppb	References
2,3,3-Trimethyl-1-butene	C_7H_{14}	98	—	93
2,3-Dimethyl-1-pentene	C_7H_{14}	98	—	93
2,4-Dimethyl-2-pentene	C_7H_{14}	98	—	93
2-Methyl-3-hexene	C_7H_{14}	98	—	93
3-Methyl-3-hexene	C_7H_{14}	98	—	93
5-Methyl-1-hexene	C_7H_{14}	98	—	93
1-Heptene	C_7H_{14}	98	0.2–33	91–93, 96, 98, 99
2-Heptene	C_7H_{14}	98	—	93
1,2-Dimethylcyclopentanes (*cis-*, *trans-*)	C_7H_{14}	98	0.4–24	92, 99
1,3-Dimethylcyclopentanes (*cis-*, *trans-*)	C_7H_{14}	98	0.2	58, 92, 93, 99
Methylcyclohexane	C_7H_{14}	98	0.4–103	56, 58, 89, 91–93, 95–100
2,2-Dimethylpentane	C_7H_{16}	100	0.2–11.2	93, 100
2,3-Dimethylpentane	C_7H_{16}	100	0.2–11.2	93, 100
2,4-Dimethylpentane	C_7H_{16}	100	0.2–25.6	56, 58, 87, 91–93, 96–100
3,3-Dimethylpentane	C_7H_{16}	100	0.2–13.7	91–93, 96–99
3-Ethylpentane	C_7H_{16}	100	0.1–4.9	58, 93, 95
2-Methylhexane	C_7H_{16}	100	1.1–40	56, 58, 85, 87, 92, 93, 95–100
3-Methylhexane	C_7H_{16}	100	1.1–68	56, 58, 87, 91–93, 95–100
Heptane	C_7H_{16}	100	0.2–95	56, 58, 84, 85, 87–100
Styrene	C_8H_8	104	0.2–22.6	87–93, 97
Ethylbenzene	C_8H_{10}	106	0.06–37.6	58, 84–93, 95–100
o-Xylene	C_8H_8	106	1.7–39.1	56, 58, 84–100
m-Xylene	C_8H_8	106	3.2–65.8	56, 58, 84–100
p-Xylene	C_8H_8	106	2.1–43.6	56, 58, 84–100
1-Octene	C_8H_{16}	112	0.04–47.6	88, 91, 92, 96–98
1,2,3-Trimethylcyclopentane	C_8H_{16}	112	0.4–20.3	92, 96–100
1,2,4-Trimethylcyclopentane	C_8H_{16}	112	0.4–28.3	92, 96–99
1,2-Dimethylcyclohexane (*cis-*, *trans-*)	C_8H_{16}	112	0.4–28.1	92, 93, 97–100
1,3-Dimethylcyclohexane (*cis-*, *trans-*)	C_8H_{16}	112	0.2–10.6	92, 96–98
1,4-Dimethylcyclohexane	C_8H_{16}	112	0.4–59.6	92, 96–98
Ethylcyclohexane	C_8H_{16}	112	0.2–31.9	91, 92, 96, 100
2,2,3-Trimethylpentane	C_8H_{18}	114	0.02–0.1	99
2,2,4-Trimethylpentane	C_8H_{18}	114	0.04–1.6	56, 58, 85, 93, 100
2,3,3-Trimethylpentane	C_8H_{18}	114	—	93, 97, 100
2,3,4-Trimethylpentane	C_8H_{18}	114	—	93, 100
2-Methyl-3-ethylpentane	C_8H_{18}	114	—	93, 100
3-Methyl-3-ethylpentane	C_8H_{18}	114	—	93, 100
2,2-Dimethylhexane	C_8H_{18}	114	0.04–3	99

Table 2.7. (*Continued*)

Compound	Formula	Mol. wt.	Concentr., ppb	References
2,3-Dimethylhexane	C_8H_{18}	114	0.06–4.3	59, 93, 100
2,4-Dimethylhexane	C_8H_{18}	114	0.02–1.0	58, 90, 93, 95, 100
2,5-Dimethylhexane	C_8H_{18}	114	0.02–1.0	58, 90, 93, 100
3,3-Dimethylhexane	C_8H_{18}	114	—	93
3,4-Dimethylhexane	C_8H_{18}	114	—	100
2-Methylheptane	C_8H_{18}	114	0.4–52	85, 87, 91, 92, 96–100
3-Methylheptane	C_8H_{18}	114	1.0–41	58, 91–93, 96–100
4-Methylheptane	C_8H_{18}	114	0.2–10.6	91–93, 96–100
3-Ethylhexane	C_8H_{18}	114	—	93
Octane	C_8H_{18}	114	1.2–78	56, 58, 84, 85, 87, 91–99
Indene	C_9H_8	116	—	88, 91–94
Indane	C_9H_{10}	118	—	88, 91–94
α-Methylstyrene	C_9H_{10}	118	0.2–4.0	91, 92, 97
1-Methyl-2-ethylbenzene	C_9H_{12}	120	0.6–19.2	56, 58, 84, 85, 87, 88, 91–93, 95–100
1-Methyl-3-ethylbenzene	C_9H_{12}	120	0.9–22.4	56, 58, 84, 85, 87, 88, 91–93, 95–100
1-Methyl-4-ethylbenzene	C_9H_{12}	120	0.6–19.6	56, 58, 86–88, 91–93, 95–100
1,2,3-Trimethylbenzene	C_9H_{12}	120	0.2–9.5	58, 85, 87–89, 91–93, 95–100
1,2,4-Trimethylbenzene	C_9H_{12}	120	0.4–31.4	56, 58, 85–89, 91–93, 95–100
1,3,4-Trimethylbenzene	C_9H_{12}	120	0.6–19.1	56, 58, 85–89, 91–93 95–100
Isopropylbenzene	C_9H_{12}	120	0.2–12.0	89, 91–93, 96–98, 100
Propylbenzene	C_9H_{12}	120	0.4–14.2	56, 58, 91–93, 95–100
1-Nonene	C_9H_{18}	126	0.2–21.7	88, 91, 92, 96–98
4-Nonene	C_9H_{18}	126	—	93
1,1,3,4-Tetramethylcyclo- pentane	C_9H1_8	126	—	100
Trimethylcyclohexanes (3 isomeres)	C_9H_{18}	126	0.4–24	91, 92, 96–99
Methylethylcyclohexanes (3 isomeres)	C_9H_{18}	126	—	91, 92
2,5-Dimethylheptane	C_9H_{20}	128	0.2–12.3	92, 93, 96, 98, 99
2,6-Dimethylheptane	C_9H_{20}	128	0.2–11.6	92, 96, 98, 99
3,4-Dimethylheptane	C_9H_{20}	128	0.02–0.2	99
2-Methyloctane	C_9H_{20}	128	0.2–14	87, 91–93, 96–100
3-Methyloctane	C_9H_{20}	128	0.5–17	91–93, 96–100
4-Methyloctane	C_9H_{20}	128	0.2–11.2	93, 95, 96, 98
Nonane	C_9H_{20}	128	0.7–40	56, 58, 84, 85, 87, 88, 91–93, 95–100
Naphtalene	$C_{10}H_8$	128	0.1–2.1	58, 85, 87, 88, 91–95

Table 2.7. (*Continued*)

Compound	Formula	Mol. wt.	Concentr., ppb	References
Tetrahydronaphtalene	$C_{10}H_{12}$	132	—	88
1,2,3,4-Tetramethylbenzene	$C_{10}H_{14}$	134	—	85, 88, 89, 92, 93
1,2,3,5-Tetramethylbenzene	$C_{10}H_{14}$	134	0.3–0.9	85, 88, 89, 92, 93
1,2,4,5-Tetramethylbenzene	$C_{10}H_{14}$	134	0.3–0.7	85, 88, 89, 92
1,2-Dimethyl-3-ethylbenzene	$C_{10}H_{14}$	134	0.01–0.6	99
1,2-Dimethyl-4-ethylbenzene	$C_{10}H_{14}$	134	0.1–0.7	99
1,3-Dimethyl-4-ethylbenzene	$C_{10}H_{14}$	134	0.05–0.4	99
1,3-Dimethyl-5-ethylbenzene	$C_{10}H_{14}$	134	0.03–0.8	92, 93, 99
1,4-Dimethyl-2-ethylbenzene	$C_{10}H_{14}$	134	0.03–0.2	99
o-Cymene	$C_{10}H_{14}$	134	0.2–4.8	85, 93
m-Cymene	$C_{10}H_{14}$	134	0.2–9.3	85, 92, 93, 96–98
p-Cymene	$C_{10}H_{14}$	134	—	85, 92, 96–98
1,2-Diethylbenzene	$C_{10}H_{14}$	134	—	87, 93, 95
1,3-Diethylbenzene	$C_{10}H_{14}$	134	—	93
1-Methyl-2-propylbenzene	$C_{10}H_{14}$	134	0.02–0.05	93, 95, 99
1-Methyl-3-propylbenzene	$C_{10}H_{14}$	134	0.02–0.8	85, 92, 99
1-Methyl-4-propylbenzene	$C_{10}H_{14}$	134	0.05–0.3	92, 93, 99
Isobutylbenzene	$C_{10}H_{14}$	134	—	87
sec-Butylbenzene	$C_{10}H_{14}$	134	—	85, 88, 93
t-Butylbenzene	$C_{10}H_{14}$	134	—	88, 93
Butylbenzene	$C_{10}H_{14}$	134	0.2–5.2	85, 87, 88, 91–93
α-Pinene	$C_{10}H_{16}$	136	—	90–92, 95
β-Pinene	$C_{10}H_{16}$	136	—	90, 92
Camphene	$C_{10}H_{16}$	136	—	91, 92
3-Carene	$C_{10}H_{16}$	136	—	92
Limonene	$C_{10}H_{16}$	136	0–5.8	85, 87, 90–92, 95
Decaline	$C_{10}H_{18}$	138	—	88
1-Decene	$C_{10}H_{20}$	140	0.16–7.7	91, 92, 96–99
t-Butylcyclohexane	$C_{10}H_{20}$	140	—	93
Butylcyclohexane	$C_{10}H_{20}$	140	0.2–8.8	91, 92, 96–98
2-Methylnonane	$C_{10}H_{22}$	142	0.2–8.3	91, 92, 96–98
3-Methylnonane	$C_{10}H_{22}$	142	0.1–5.1	91, 92, 96–99
4-Methylnonane	$C_{10}H_{22}$	142	0.2–5.4	91, 92, 96–99
5-Methylnonane	$C_{10}H_{22}$	142	—	92
Decane	$C_{10}H_{22}$	142	0.6–24.4	56, 58, 84–88, 91–93, 95–100
1-Methylnaphtalene	$C_{11}H_{10}$	142	—	85, 88
2-Methylnaphtalene	$C_{11}H_{10}$	142	—	85, 88
Amylbenzene	$C_{11}H_{16}$	148	—	91, 92
Pentamethylbenzene	$C_{11}H_{16}$	148	—	88
1-Undecene	$C_{11}H_{22}$	154	0.1–9.2	92, 97, 98
Acenaphtene	$C_{12}H_{10}$	154	—	92, 96–98
2-Methyldecane	$C_{11}H_{24}$	156	0.1–4.7	92, 96–98
3-Methyldecane	$C_{11}H_{24}$	156	0.2–7.5	92, 96–98

Table 2.7. (*Continued*)

Compound	Formula	Mol. wt.	Concentr., ppb	References
Undecane	$C_{11}H_{24}$	156	0.2–12.4	58, 84–88, 92, 93, 95–99
1,4-Dimethylnaphtalene	$C_{12}H_{12}$	156	—	88
1,6-Dimethylnaphtalene	$C_{12}H_{12}$	156	—	85, 88
1,8-Dimethylnaphtalene	$C_{12}H_{12}$	156	—	85
2,3-Dimethylnaphtalene	$C_{12}H_{12}$	156	—	88
2,6-Dimethylnaphtalene	$C_{12}H_{12}$	156	—	85
1-Ethylnaphtalene	$C_{12}H_{12}$	156	—	88
2-Ethylnaphtalene	$C_{12}H_{12}$	156	—	88
Hexylbenzene	$C_{12}H_{18}$	162	—	92
Hexamethylbenzene	$C_{12}H_{18}$	162	—	88
Fluorene	$C_{13}H_{10}$	166	—	85, 88
Diphenylmethane	$C_{13}H_{12}$	168	—	88
1-Dodecene	$C_{12}H_{24}$	168	—	88, 91, 92
Dodecane	$C_{12}H_{26}$	170	0.2–1.3	58, 84, 85, 87, 88, 92, 93, 95, 99
1-Tridecene	$C_{13}H_{26}$	182	—	88, 91, 92
Tridecane	$C_{13}H_{22}$	184	—	85, 87, 88, 91–93, 95
Tetradecane	$C_{14}H_{30}$	198	0.9–2.5	84, 85, 87, 88
Pentadecane	$C_{15}H_{32}$	212	—	85, 87, 88
Hexadecane	$C_{16}H_{34}$	226	0.2–1.3	85, 87, 88
Heptadecane	$C_{17}H_{36}$	240	—	85, 88
Octadecane	$C_{18}H_{40}$	254	—	85
Nonadecane	$C_{19}H_{40}$	268	—	85

list which includes 193 compounds, whenever possible the concentration range in which a given hydrocarbon has been found in the atmosphere of various cities is also reported. Sample collection was usually carried out in non-industrial regions, and therefore this list characterizes the general composition of hydrocarbons in the air of the business part and the blocks of dwelling house of these cities. This group of organic components includes alkanes, napthenic, olefinic, acetylenic and aromatic hydrocarbons.

The study of this list of volatile hydrocarbons indicates, primarily, that almost all of them have been repeatedly found in the atmosphere of various cities. Hence, the comparison of lists of substances found in the atmosphere of cities suggests that the qualitative compositions of volatile hydrocarbons in the air basins of modern cities are very similar regardless of the geographical location and the size of these cities.

Although the qualitative composition of pollutants is relatively uniform, their absolute concentrations undergo considerable variations with time even in the atmosphere of one city. The highest hydrocarbon concentrations are observed in the morning hours and in early evening, and the minimum occurs at night-time. During one week, the highest hydrocarbon content is observed from Tuesday to Friday, and the minimum is found on Saturday and Sunday.

Table 2.8. Average concentration (ppb) of aromatic C_6–C_9 hydrocarbons in some urban atmospheres

Hydrocarbon	Los Angeles [96]	Houston [87]	The Hague [100]		West Berlin [56]		Johannesburg [93]	Leningrad [96]	Havana [99]
			Prinsengracht	Maurits-kade	Dahlem	Steglitz			
Benzene	6.0 ± 4.6	1.3–15.0	4.6	13.1	2.2	14.6	3.4	29.9 ± 22.0	3.1
Toluene	11.7 ± 9.0	0.3–9.7	13.1	22.7	2.8	24.8	10.3	27.0 ± 15.6	8.8
Ethylbenzene	2.3 ± 4.4	3.1–4.5	1.2	5.0	0.7	5.7	1.4	4.2 ± 2.5	1.1
p-Xylene	4.6 ± 6.1[a]	2.1–3.4	5.1[a]	19.6[a]	1.3[a]	11.2[a]	4.5[a]	4.9 ± 2.3	3.4[a]
m-Xylene	—	5.4–7.8	—	—	—	—	—	12.7 ± 2.3	—
o-Xylene	1.9 ± 1.9	3.0–4.8	1.3	4.9	0.5	5.1	1.6	6.5 ± 5.7	1.1
n-Propylbenzene	—	1.5–4.0	—	—	1.9[b]	12.5[b]	0.4	1.8 ± 0.7	0.6
Cumene	—	—	—	—	—	—	—	1.7 ± 0.9	—
m,p-Ethyltoluene	1.5 ± 1.5[c]	—	2.4[c]	9.8[c]	—	—	1.6	3.4 ± 2.0	1.7
1,3,5-Trimethylbenzene	0.4 ± 0.7	—	1.2	1.9	—	—	0.6	1.5 ± 0.9	0.4
1,2,4-Trimethylbenzene	1.9 ± 2.4	—	4.4	10.1	—	—	1.6	3.4 ± 1.9	1.1

[a]Sum of p- and m-isomers, [b]Sum of C_9H_{12} isomers, [c]only p-isomer

Both the similarity of composition of hydrocarbon components and the peculiar diurnal and weekly variations in concentrations in the air of cities located on different continents doubtless result from a common source of their emission into the atmosphere. Almost all the compounds listed in Table 2.7 are contained in car engine fuel. All the rest, except terpenes (which are of biogenic origin), are formed as a result of destructive processes in car engines and are emitted into the atmosphere with the exhaust gases.

Many researchers focus great attention on the determination of the content of aromatic hydrocarbons in the urban atmosphere because benzene and its homologues are toxic substances. The study of the behavior of alkyl benzenes in smog chambers has shown that in the presence of nitrogen oxides they readily take part in reactions leading to the formation of aerosols, ozone and other secondary pollutants of urban air. Di- and trialkylbenzenes are particularly active in photochemical reactions (see chap. 5).

Table 2.8 gives data on the content of aromatic hydrocarbons in the air of some cities of America, Africa and Europe. These values usually represent concentrations obtained by averaging the results of many determinations. The table also lists the values of standard deviations or the concentration range in which a given hydrocarbon has been found in the atmosphere. Apart from the literature data, this table lists some of the results obtained by the author in the investigations of the composition of pollutants in the air basin of Leningrad in 1976–1980. In these years the relative content of aromatic hydrocarbons in Leningrad's atmosphere was comparatively stable and made up $34 \pm 4\%$ of the sum of the C_4–C_{12} hydrocarbons. Similar values of the relative content of benzene and its homologues are also characteristic of eight other large industrial

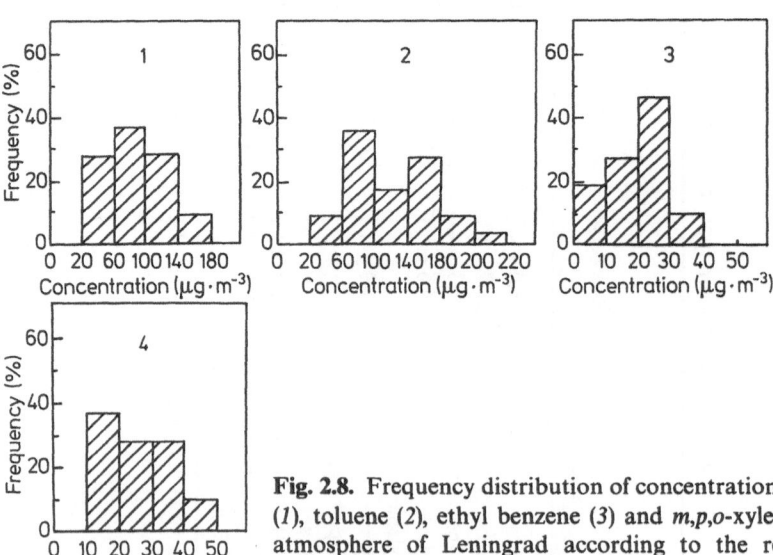

Fig. 2.8. Frequency distribution of concentrations of benzene (1), toluene (2), ethyl benzene (3) and m,p,o-xylene (4) in the atmosphere of Leningrad according to the results of 50 analyses carried out in 1980 [101]

cities of the USSR [101]. Hence, volatile aromatic hydrocarbons are one of the major fractions of the organic component of the urban atmosphere. The histograms in Fig. 2.8 give an idea about the frequency distribution of concentrations of six compounds which usually make up over 80% of the total amount of aromatic hydrocarbons in the Leningrad air.

2.2 Oxygen-, Nitrogen- and Sulfur-Containing Components

In addition to hydrocarbons, their numerous derivatives containing different functional groups are present in the atmospheric air. They are very reactive, and hence their lifetime in the atmosphere is relatively short and generally does not exceed a few hours and sometimes even minutes. Therefore these shortlived compounds can often be detected only in the immediate vicinity of the sources from which they are emitted into the atmosphere. However, as compared to most hydrocarbons, their derivatives the molecules of which contain oxygen, nitrogen and sulfur atoms exhibit high physiological activity and profoundly affect the quality of human environment.

Table 2.9 lists several tens of compounds found in the open atmosphere of nonurban areas and in the air of blocks of apartment houses of cities in concentrations not lower than 0.01 ppb. The total amount of volatile hydrocarbon derivatives present in the atmosphere constantly or even from time to time is probably very large. It may be assumed that apart from the components of natural origin emitted by living organisms or decomposed plant and animal tissues, virtually all the compounds which are intermediate and final products of industrial organic synthesis are emitted into the atmosphere. The data available in the literature on the content of some compounds of those listed in Table 2.9 and being the most important classes of hydrocarbon derivatives will be discussed below.

2.2.1 Carbonyl Compounds

Just like hydrocarbons, carbonyl compounds are permanent components of the Earth's atmosphere. However, in contrast to the former the sources of which are located on the Earth's surface and are related to the vital activity of various organisms or the industrial activities of man, many carbonyl compounds are also formed directly in the atmosphere as a result of some photochemical processes. This refers primarily to the simplest representative of this class: formaldehyde (CH_2O).

The presence of formaldehyde in the atmospheric air was first established as early as the 1930s by Dhar N.R. and Ram A. [102a]. Even the first determinations carried out in the late 1940s showed that it is present in the atmosphere over a wide concentration range: in the relatively pure air transported to Germany's shores from the Baltic sea, its concentration was 0.7 to 2 ppb, whereas in the continental densely populated areas of Europe it attained 50 ppb [102b].

Table 2.9. Oxygen-, nitrogen- and sulfur-containing compounds detected in the asmosphere

Compound	Formula	Mol. wt.	Concentrations, ppb Background	Concentrations, ppb In the urban air	References
Formaldehyde	CH_2O	30	0.02–11	1–160	103–115
Methyl amine	CH_5N	31	$(4.2 \pm 0.1) \cdot 10^{-4}$	—	116
Methanol	CH_4O	32	—	0.3–59	12, 56, 90
Acetonitrile	C_2H_3N	41	0.02–12	7	117–121
Ethylene oxide	C_2H_4O	44	—	—	87
Acetaldehyde	C_2H_4O	44	0.1–3	1–56	12, 56, 90, 93, 94, 121
Dimethylamine	C_2H_7N	45	$(5.4 \pm 0.1) \cdot 10^{-3}$	—	90, 116
Ethylamine	C_2H_7N	45	—	—	90
Formic acid	CH_2O_2	46	0.2–1.5	0.4–19	122–125
Ethanol	C_2H_6O	46	—	1.0–69	12, 90, 91, 93
Methyl mercaptan	CH_4S	48	$1 \cdot 10^{-3}$	—	89, 129
Acrolein	C_3H_4O	56	—	0.1–14	94
Acetone	$C_3{}^H{}_6O$	58	0.1–10	1–120	12, 58, 89–92, 94, 95, 120
Propanal	C_3H_6O	58	—	0.5–37	110
Trimethylamine	C_3H_9N	59	$(2.3 \pm 0.1) \cdot 10^{-3}$	—	116
Acetic acid	$C_2H_4O_2$	60	0.2–1.0	0.2–6	122–127
Propanol	C_3H_8O	60	—	0.01–0.03	121
2-Propanol	C_3H_8O	60	—	—	90
Methyl nitrate	CH_3NO_2	61	—	5	
Dimethyl sulfide	C_2H_6S	62	$(0.7–10) \cdot 10^{-3}$	0.04–0.5	128–132
Isobutanal	C_4H_8O	72	—	—	93
Butanal	C_4H_8O	72	—	0–8	93, 110
2-Butanone	C_4H_8O	72	0.5 ± 0.2	0.03–15	110–112, 121
1-Butanol	$C_4H_{10}O$	74	0.06 ± 0.03	0.06–0.2	12, 89, 90, 121
t-Butanol	$C_4H_{10}O$	74	—	0.12 ± 0.06	121
Diethyl ether	$C_4H_{10}O$	74	—	—	89, 90, 92
Methyl acetate	$C_3H_6O_2$	74	—	—	90
Dimethylnitrosamine	$C_2H_6N_2O_2$	74	—	$(0.3–11) \cdot 10^{-3}$	132
Ethyl nitrite	$C_2H_5NO_2$	75	—	—	133
2-Methylfuran	C_5H_6O	82	—	—	91, 93
3-Methylfuran	C_5H_6O	82	—	—	134
2-Pentanone	$C_5H_{10}O$	86	—	—	93
3-Pentanone	$C_5H_{10}O$	86	—	—	89
1,4-Dioxane	$C_4H_8O_2$	88	—	—	91, 92, 95
Dimethyl disulfide	$C_2H_6S_2$	94	—	—	133
Phenol	C_6H_6O	94	—	0.5–2.5	94
Ethylfuran	C_6H_8O	96	—	—	90
Isopropyl acetate	$C_5H_{10}O_2$	102	—	—	135
Benzaldehyde	C_7H_6O	106	—	0.4–2	85, 87, 90, 94, 95, 110
o-Cresol	C_7H_8O	108	—	0.07–0.35	94
3-Heptanone	$C_7H_{14}O$	114	—	—	89
Butyl acetate	$C_6H_{12}O_2$	116	—	—	90, 91, 95
Benzofuran	C_8H_6O	118	—	—	85

Table 2.9. (*Continued*)

Compound	Formula	Mol. wt.	Concentrations, ppb Backgro- und	Concentrations, ppb In the urban air	References
o-Toluyl aldehyde	C_8H_8O	120	—	—	85, 87, 94
m-Toluyl aldehyde	C_8H_8O	120	—	0.5–0.6	85, 87, 94
p-Toluyl aldehyde	C_8H_8O	120	—	—	85, 87, 94
Acetophenone	C_8H_8O	120	—	—	85, 87, 94
Peroxyacetyl nitrate	$C_2H_3NO_5$	121	—	0.003–66	86, 122
2-Hydroxybenzaldehyde	$C_7H_6O_2$	122	—	—	135
4-Hydroxybenzaldehyde	$C_7H_6O_2$	122	—	—	135
2-Hydroxybenzyl alcohol	$C_7H_8O_2$	124	—	—	135
4-Hydroxybenzyl alcohol	$C_7H_8O_2$	124	—	—	135
Isoamyl acetate	$C_7H_{14}O_2$	130	—	—	90
Methylacetophenone	$C_9H_{16}O$	134	—	—	87
Peroxypropionyl nitrate	$C_3H_5NO_5$	135	0.008–0.012	0.03–2.7	86, 140, 141
Benzothiazole	C_7H_5NS	135	—	—	85
Peroxybutyryl nitrate	$C_4H_7NO_5$	149	—	—	112

Table 2.10 lists data on the background concentrations of formaldehyde in the air over the seas and on its content over the continents in rural and urban areas. The table shows that in the middle latitudes of the Northern Hemisphere, the background concentrations of formaldehyde in the air masses arriving from sea regions do not exceed 0.4 ppb. Some authors have reported the dependence of the CH_2O content on weather conditions: the maximum concentrations were usually observed during warm sunny weather, whereas during long periods of rainy weather and after rainfall they decreased markedly.

At present, little information is available about the formaldehyde content in air in low latitudes. However, it seems likely that near the tropics its content is slightly higher than in higher latitudes of both hemispheres. The highest concentrations ranging from 3 to 11 ppb have been detected by Japanese researchers [106] in sample collection on board a research ship in the regions of the Philippine Islands and Indonesia. An increase in the formaldehyde content to 3.7 to 8.2 ppb has also been reported when the ship moved along the Western shore of Australia (20°S–36°S).

It may be suggested that the increase in the formaldehyde content in low latitudes is due to a high intensity of photochemical processes in the atmosphere caused by a high level of solar radiation. The unusually high concentration of CH_2O in the air of equatorial and tropical regions near the Philippine Islands and the Malay Archipelago can result from the presence of large biogenic sources of hydrocarbons being formaldehyde precursors in photochemical reactions (see chap. 5).

Table 2.10. Formaldehyde content in atmospheric air

Location of sample collection	Concentrations, ppb		
	Average	Concentration range	Refs.
Sea air			
Federal Republic Germany, Northern shore, 54.5 °N	—	0.02–0.4	103
Ireland, Western shore, 52.5 °N	—	0.07–0.3	103
The same location	—	0.1–0.42	104
Central Atlantic, 10 °N–10 °S	—	0.1–0.3	105
Pacific Ocean:			
35 °–25 °N	—	0.7–2.2	106
Enewetak Atoll, 11 °24′	0.40 ± 0.20	0.31–0.57	107
South China Sea, 10°–0.1°N	—	3–11	106
Indian Ocean:			
9 °–18 °S	—	1.5–3	106
12 °–47 °S	—	0.8–3	106
Air of non-urban areas			
U.S.A: Whiteface Mnt, New York	—	0.8–0.26	108
Long Island, New York	7.5	—	109
F.R.G.: Deuselbach, April 1978	—	0.03–0.9	103
Jülich, 1978	—	0.1–6.5	103
Jülich, 1979–1980	1.3 ± 0.7	0.3–4.5	102
Jülich, 1980	—	0.2–2.5	104
Jülich, winter 1980	—	0.08–0.45	105
Japan, Mnt. Norikura (2700 m)	—	0.7–3	106
Urban air			
U.S.A.: Schenectady, New York	—	1–31	108
Los Angeles, California	27.5	—	110
Los Angeles, California	—	18–70	111
Clermon, California	—	8.2–47.8	112
Riverside, California	19.0 ± 7.6	41.0[a]	113
Downey, California	15.5 ± 11.9	67.7[a]	113
St. Louis, Missouri	11.3 ± 4.5	18.7[a]	113
Denver, Colorado	12.3 ± 5.9	28.7[a]	113
Chicago, Illinois	11.3 ± 3.8	28.5[a]	113
Pittsburgh, Pennsylvania	18.7 ± 6.7	28.5[a]	113
State Island, New York	14.3 ± 9.1	45.9[a]	113
Bayonne, New Jersey	10.0	—	114
Newark, New Jersey	13.0	—	114
Japan, Tokyo	—	20–35	115

[a]Maximal concentration

At present, there are virtually no data on the CH_2O content in the atmosphere over the continents in regions where there is no anthropogenic effect. In the air of rural areas near the industrial centers, the variations in the amount of formaldehyde with time are very great and attain one or two orders of magnitude. Thus, in one of the areas of the German Federal Republic (Northern Rheinland Westfalia) according to the data of 64 analyses, the concentration of formaldehyde varied from 0.3 to 4.5 ppb but most often was 0.82 ± 0.07 ppb [103]. The same authors determined the CH_2O content at different altitudes in the autumn of 1979 over the territory of the German Federal Republic. The analysis showed that at altitudes of 0.4 to 3 km the concentrations were 0.2 to 0.5 ppb, whereas over the temperature inversion layer they were lower than 0.07 ppb.

The highest formaldehyde concentrations are detected in urban air [110–115]. The statistical analysis of data on the measurement of CH_2O content in the air basin of four relatively small towns in the east of the USA [114] shows that traffic intensity is the determining factor. A direct relationship between air pollution by formaldehyde and the intensity of traffic is confirmed by a peculiar diurnal and weekly concentration trend.

The maximum CH_2O content is observed in the morning hours and in early evening, and the minimum content is detected at night. During the week the highest concentrations fall on work-days, whereas on Saturday and Sunday they decrease considerably. The same fluctuations are also characteristic of the content of carbon oxide and hydrocarbons the main source of which is automobile transport. The authors of ref. [114] have also reported a direct dependence of formaldehyde concentration on the level of solar radiation and hence on the intensity of photochemical processes in the urban atmosphere containing large amounts of hydrocarbons.

Apart from formaldehyde, other aliphatic and aromatic carbonyl compounds are emitted into the urban atmosphere from various sources. According to the data in ref. [106], the total content of aldehydes in the air of Tokyo attains 30 ppb, whereas their concentrations in the atmosphere of Los Angeles at the beginning of the 1980s varied from 21 to 241 ppb [111]. Acetaldehyde is present in the urban air most often and in greatest amounts. It has been reported that its concentrations in the air basins of some Japanese cities were 2 to 8 ppb [115], in that of West Berlin they varied from 1 to 12.2 ppb [56] and in Paris they ranged from 4.6 to 29 ppb [94].

High concentrations of formaldehyde homologues were detected in the air of metropolian Los Angeles in smog situations. Thus, in July–October 1980 the concentration of acetaldehyde attained 56 ppb and those of propanal and butanal were 37 and 8 ppb, respectively [112]. According to the data of Altshuller A.P. and McPherson S.P. (1968), the concentrations of acrolein in the air of Los Angeles during smog were on the average 7 ppb and the maximum concentrations attained 14 ppb.

Aromatic aldehydes are usually present in smaller amounts. Benzaldehyde concentrations in the air of Los Angeles generally do not exceed 2 ppb [110]. In the air of Paris, Hagemann R. et al. have found benzaldehyde and tolylaldehyde at concentrations of 0.4–1.3 ppb and 0.04–0.6 ppb, respectively [94].

Few data are available in the literature on ketone concentrations in the urban atmosphere. Lahman E. et al. [56] found acetone in various areas of West Berlin in the period from January to July 1977 with concentrations from 1.7 to 118 ppb. Average acetone concentrations in the air of the streets of Göteborg were 15.8 ppb [136]. 2-Butanone is also present in considerable amounts sometimes attaining 15 ppb [112]. These ketones are constant components of the urban air and were detected in all the cities of the USSR investigated [91, 92].

Aliphatic aldehydes and ketones are also present in the atmosphere of rural areas. Their amount depends, first, on the distance from the cities and superhighways. It has been observed, for example, that the total aldehyde concentration in the air of a rural area in Panama varies from less than 0.1 to 3 ppb. Similar values have been detected in Japanese mountains [106], whereas in the central states of the USA they have attained 10 ppb.

In the air masses of the Arctic regions, only small amounts of acetaldehyde (from less than 0.1 to 0.3 ppb) and acetone (0.3 to 2.9 ppb) have been detected [12]. However, in remote regions in the jungles of the Amazon basin, the acetone content in air during the dry season of 1970 was found to be unusually high and attained 240 ppb [83]. It has been suggested [83] that these high acetone concentrations are a result of photochemical transformations of a number of organic compounds emitted by plants. However, in our opinion it cannot be ruled out that large amounts of acetone are emitted into the atmosphere by the tropical vegetation itself.

In recent years it has been established that carbonyl compounds are present not only in the overground layer of air but also in the upper troposphere and even in the stratosphere, although their concentrations are low. The authors of ref. [137] have recorded in the upper troposphere over the German Federal Republic the average acetone concentrations of 120 ppt. Above the tropopause, a drastic decrease in average concentrations to 31 ppt has been recorded. Acetone content at an altitude of 11,300 m did not exceed 2 ppt [120]. These carbonyl compounds are probably formed directly in the upper layers of the atmosphere as a result of photochemical oxidation of ethane and propane.

2.2.2   Alcohols and Carboxylic Acids

The data on the contents of alcohols and carboxylic acids both in the background atmosphere and in the urban air are very scarce. According to the report of the authors of ref. [12], the total content of methyl and ethyl alcohol in the Arctic masses of air in Alaska in the summer of 1967 did not exceed 1.2 ppb. However, in the same period the concentrations of n-butyl alcohol were unusually high and varied from 34 to 445 ppb (on average 190 ppb), i.e. they exceeded the content of any other of the C_2–C_6 compounds detected there by more than two orders of magnitude. Some microbiological processes occurring in the upper layers of the tundra soil were found to be responsible for this very high n-butanol content.

According to the report in ref. [56], lower alcohols are continuously emitted into the atmosphere of large cities. The content of methyl alcohol in the air of West Berlin in 1977 varied from 3 to 59 ppb at an average concentration of about 14 ppb. The concentrations of ethyl alcohol ranged from 1.1 to 68.2 ppb.

At the beginning of the 1980s, Snider and Dawson [118] determined the content of C_1–C_4 alcohols in two rural areas and in a small town in the southwest of the USA (Arizona). According to their data, at all three locations methanol was present in the largest amount. Its concentration in the town of Tucson was 5.5 ± 1.8 ppb, and in rural areas it was 1.8 ± 0.8 ppb. In the homologous series of alcohols, concentrations decreased rapidly: on passing from methanol to ethanol and butyl alcohols the average concentrations decreased by factors of 2 and more than 25, respectively.

Very few data have been published on the content of aromatic alcohols in the urban air. Hagemann R. et al. [94] have detected phenol and O-cresol in the air of Paris at concentrations of 0.54–2 ppb and 0.07 to 0.35 ppb, respectively.

The simplest carboxylic acids appear in the urban air mainly as a result of atmospheric photochemical processes, as is indicated by the existence of a correlation between their content and that of ozone and other photooxidants. The highest concentration of formic acid detected in critical smog situation in Los Angels suburbs in 1973 and 1978 was 39 ppb. A much lower content of formic acid is characteristic of the atmosphere of non-industrial towns. For example, in the town of Tucson (Arizona) mentioned above its concentration was at the level of 5 ppb and did not exceed 7 ppb [122]. The maximum amounts of acetic acid in this town were approximately 16 ppb but more often ranged from 3 to 9 ppb. In the suburbs, the content of formic and acetic acids was 0.8 to 3.1 and 0.5 to 2.5 ppb, respectively. Dawson et al. [122] reported that the minimum concentrations of acids are observed after precipitation. This fact is due to the high solubility of acids in water. Kawamura K. and Kaplan F.H. [123] investigated the acid content in rain water and fog samples collected in Los Angeles in 1982–1983. According to their data, the concentrations of the C_1–C_6 acids vary from 51 to 95.2 μmole$\,l^{-1}$. Furthermore, the contribution of the first two members of the homologous series usually provides more than 98% of this amount. Considerable concentrations of acids (up to 145 μmole$\,l^{-1}$) were also discovered in fog drops, and the content of formic acid (81 to 94 μmole$\,l^{-1}$) was twice that of acetic acid.

Keen W.C. et al. who carried out analyses of the composition of precipitation in various regions of the world from 1979 to the middle of the 1980s [126] have concluded that in contrast to the previously published data, the main contribution to the acidity of rains in many of these regions is provided not by mineral but by organic acids. Their contribution to the acidity of precipitation over the areas remote from the cities is particularly high. For example, the mean pH value of rain water over the Amazonia was 4.81, 49 to 65% of which was due to the presence of $HCOOH$ and CH_3COOH. In the industrial regions of North America, these acids provide not less than 16 to 20% of the acidity of precipitation [125].

At present, it has been established that low molecular weight acids are present throughout the troposphere. The concentrations of HCOOH at altitudes of 8 and 10 km over the USA are 0.6 and 0.4 ppb, respectively [124]. Just as the lower carbonyl compounds, acids are mainly the products of photochemical oxidation of hydrocarbons.

2.2.3   Nitrogen- and Sulfur-Containing Components

Nitrogen-containing organic substances are usually present in the atmosphere in very small amounts. However, two groups of compounds, N-nitrosamines and peroxyacyl nitrates, attract great attention.

N-nitrosamines belong to carcinogenic substances, and one of the most active compounds in this respect is the simplest representative of this series: dimethyl nitrosamine. It was first detected in the atmospheric air in the middle of the 1970s. Fine D.H. et al. [132] observed it in all samples collected in Baltimore in the winter of 1975 at concentrations from 0.016 to 0.76 ppb. The greatest amounts of dimethylnitrosamine were detected in air near the industrial enterprises manufacturing or using as intermediate products alkylamines and dialkyl hydrazines. These researchers reported the absence of correlation between the concentration and the time of the day, the air temperature, humidity and other meteorological factors. In contrast, Chuong et al. [138] reported that in summer months the dimethylnitrosamine content in the air of Paris increased at night and considerably decreased in the day-time, particularly in the period of the maximum solar radiation. This fact indicates that nitrosamine decomposes under the influence of sunlight or chemical agents present in the atmosphere in the day-time in the greater amounts. In the atmosphere of Paris, dimethylnitrosamine has been detected only in 25% of the samples at concentrations of about 0.2 ppb.

The content of peroxyacylnitrates in air also depends on the level of solar radiation but this dependence is of a different type: the maximum concentrations are usually observed at noontime or soon after noon. The reason for this is that these compounds are secondary atmospheric pollutants formed as a result of photochemical processes. Most often the air contains peroxyacetylnitrate, $CH_3COOONO_2$(PAN), which under normal conditions is an oily liquid with a strong smell. Its vapor irritates mucuous membranes and is a potent lachrymator. PAN is a strong phytotoxin negatively affecting the development of many plants. Its appearance in the urban atmosphere is an indication of the smog situation dangerous to the health of the population.

The highest PAN concentrations, attaining 66 ppb, were recorded in the autumn of 1968 in Los Angeles. During the smog episodes in 1980, the highest PAN concentrations in the air of this city were 47 ppb [128]. In other USA cities, PAN concentrations at the end of the 1970s and in the early 1980s were on the level of 10 to 25 ppb. In the cities of Western Europe it is present in smaller amounts. For example, its maximum concentrations in the air of Bonn in the summer of 1975 were 13 ppb and those in the air of Cologne in the summer of 1979 did not exceed 5.2 ppb [139].

Recently it has been established that PAN is present in small amounts in the background air and even in the upper layers of the troposphere [140–142]. In pure continental air in the Colorado mountains (3,000 m), its concentrations usually do not exceed 170–250 ppt. Approximately the same concentrations (189–234 ppt) have been recorded at the Alert station in northern Canada [142].

The characterization of the global distribution of PAN is given in ref. [141]. As a result of numerous measurements carried out on board ship and in balloons, it was possible to establish the existence of a considerable latitudinal gradient of PAN content in pure sea air: in the 50 °–20 °N range its concentrations were 80 to 20 ppt, whereas in the equatorial region and to the south of the equator they did not exceed 10 ppt. It was found that in the upper layers of the troposphere over the Pacific Ocean the concentrations of PAN are higher than near the sea surface. In summer time they are much higher in the troposphere over the land than over the ocean.

PAN homologues are present in the atmosphere in much lower concentrations. The average concentrations of peroxypropionyl nitrate (PPN) in the air of Los Angeles in April 1979 were 721.5 ppt and in Phoenix and Oakland did not exceed 150 ppt [86]. In the urban air of the USA, the PPN/PAN ratio usually does not exceed 0.05.

Very little information is available on the concentration of aliphatic amines in the atmosphere. According to a recent report [116], in the air of the Atlantic shore of the USA (Narrangansett, Rhode Island), the average concentrations of mono-, di- and trimethylamines were 52 ± 13, 240 ± 40, and 100 ± 40 pmol m^{-3}, respectively.

Great interest has recently been aroused by such minor atmospheric component as acetonitrile because it has been detected as a component of positively charged cluster ions presumably playing an important role in the chemical processes at the altitudes of 10 to 50 km. It has been established with the aid of mass spectrometry that the principal cluster cations in the stratosphere have the following composition: $H^+(H_2O)y$ and $H^+(CH_3CN)_x(H_2O)_y$, the former predominating in the region below 35 km and the latter being more abundant in the higher regions.

The concentrations of acetonitrile in the urban atmosphere do not exceed a few ppb [117] and in the overground air of rural areas usually do not attain even 100 ppt [118, 121]. Up to the present only a few single measurements of CH_3CN in the stratosphere have been carried out. In one of these measurements, the concentrations of 3.3 to 3.4 ppt have been recorded at the altitudes of 22.6–23.7 km over France. At an altitude of 24.7 km its concentrations attained 9.5 ppt [143]. It is assumed that the main amounts of acetonitrile are not transported from the troposphere but are formed directly in the stratosphere as a result of chemical processes involving formaldehyde and HCN [144].

Volatile sulfur-containing compounds are represented mainly by methyl mercaptan, CH_3SH, dimethyl sulfide, CH_3SCH_3, and dimethyl disulfide, CH_3SSCH_3. Moreover, near some specific natural and anthropogenic sources other organic sulfur compounds: dimethyltrisulfide, thiophene and alkyl thio-

Table 2.11. Dimethyl sulfide concentrations (based on sulfur) in the sea and continental air [130, 145]

Location	Concentration, ppt	
	Range	Average
Equatorial Pacific	47.6–290	112.7
North and South Atlantic	0.7–30.8	2.8
Bahamas	3.5–469.2	85.4
Sargasso Sea	0.7–710.1	162.5
Cape Grim, Tasmania	23.8–336.8	116.9
Mt. Kleiner Feldberg, F.R.G.	3.5–14.7	6.3
Frankfurt, F.R.G.	3.5–329.1	22.4
Wiesbaden, F.R.G. (1.5 km from a chemical plant processing pulp and viscose)	24.5–444.7	138.7

phene, are also detected. Quantitative data obtained for dimethyl sulfide (DMS) are given in Table 2.11.

The characteristic feature of the global distribution of dimethyl sulfide is the fact that its concentrations in the sea air far from shore are often much higher than over the continental regions and even higher than in the moderately polluted urban atmosphere. The interpretation of this phenomenon is that on the global scale the principal sources for DMS are algae, phytoplankton and, possibly, benthos. Hence, the highest concentrations of DMS over the oceans are recorded in regions with high primary productivity. The concentration of this component in the sea air also depends on the rate of its transfer from the aqueous to the gas phase, which is determined not only by the content of dissolved DMS but also by the conditions of mixing of the upper layer of sea water. The highest concentrations are usually observed at high wind velocities [130].

The diurnal trend of DMS concentrations exhibits a maximum at night and in the morning hours and a minimum in the afternoon. This is due to the changes in both the state of the atmosphere (perturbations of stratification) and the intensity of photochemical processes. The maximum and minimum levels of DMS concentrations in the air over the equatorial regions of the Pacific Ocean differ by a factor of about 1.7 [146]. Vertical distribution also depends to a large extent on the state of the atmosphere. In the absence of convection, the concentrations of DMS decrease drastically (by up to two orders of magnitude) even in the boundary layer, whereas under the conditions of intensive convection the gradient is much less pronounced up to the altitudes of 5 to 6 km [131].

Quantitative data on other organic sulfur compounds are almost completely absent. Bingemer [145] detected methyl mercaptan in Frankfurt am Main at a level of 6.5 ppb (9.3 ngS m^{-3}) and in Wiesbaden near a chemical plant its content varied in the range of 14 to 400 ppt the average value being 126 ppt (180 ngS m^{-3}). Very high local concentrations of methyl mercaptan and its homologues (1 to 21 ppm) have been observed in the air over city landfills.

2.3   Halogenated Compounds

Halogen-containing hydrocarbon derivatives have been one of the most intensively studied groups of organic components of the atmosphere from the publication of the first papers by Rowland F.S. and Molina M.J. (1974–1975) who advanced a hypothesis about the negative consequences of emission of chloro-fluorohydrocarbons (CFC) for the protective ozone screen in the stratosphere [147, 148].

CFC's are mainly anthropogenic atmospheric pollutants (as to their natural source see Sect. 3.2). Their industrial production began in the middle of the 1930s, and virtually all the amount of these highly volatile and chemically inert compounds produced since then remained in the atmosphere, which resulted in the appearance of a considerable global background. Rowland and Molina's hypothesis is based on the assumption that there are no considerable sinks of these components in the troposhpere. As a result of atmospheric circulation, these longlived components penetrate into the upper layers of the troposphere and into the stratosphere. Photodissociation due to the action of ultraviolet solar radiation at wavelength shorter than 220 nm leads to the evolution of chlorine which interacts with ozone. This process is of a cyclic character, and the chlorine atoms play the role of the catalyst. Hence, even a relatively low amount of CFC's can result in the considerable decrease in ozone concentration in the stratosphere. This question is considered in greater detail in chap. 5.

The quantitative characteristic of the process of ozone destruction can evidently be established only with the aid of mathematical simulation the construction of which requires a large collection of data on the content, the distribution in various layers of the atmosphere, the emission rates and the photochemical dissociation of not only CFC's but also other halogen-containing pollutants.

Several tens of highly volatile halogen-containing organic compounds have been detected in the open atmosphere. They are listed in Table 2.12 with the indication of the level of this background concentrations and content in the urban air. The comparison of these figures with the data given in Table 2.5, 2.8 and 2.9 shows that halogen-containing substances even in the urban air are present in much smaller quantities (usually smaller by two or three orders of magnitude) than hydrocarbons and oxygen-containing compounds. Alkyl chlorides and chlorofluorohydrocarbons are encountered most often. Alkyl bromides are a much less abundant fraction, and iodides are represented by only one component: methyl iodide.

Apart from highly volatile components, some polychlorinated hydrocarbon derivatives widely used as the means of chemical protection of plants are constantly present in air. The data available in the literature on the content of the principal representatives of these groups of compounds in the Earth's atmosphere will be considered in the following sections.

Table 2.12. Halogen-containing compounds detected in the atmospheric air

Compound	Formula	Mol. wt.	Concentration, ppt Background	In the urban air	Ref.
Methyl chloride	CH_3Cl	50	520–640	480–7700	86, 94, 114, 149–152
Vinyl chloride	C_2H_3Cl	62	5	10–1200	56, 149
Ethyl chloride	C_2H_5Cl	64	5–19	—	149, 154
1-Chloropropane	C_3H_7Cl	78	—	—	92
2-Chloropropane	C_3H_7Cl	78	—	—	92
Dichloromethane	CH_2Cl_2	84	5–38	86–12000	86, 149, 154
Chlorodifluoro-methane	CHF_2Cl	86	2–58	—	151, 155, 156
Tetrafluoromethane	CF_4	88	70–75	70	151, 157
1-Chlorobutane	C_4H_9Cl	92	—	—	92, 133
Methyl bromide	CH_3Br	94	5–23	4–1033	86, 149, 152, 161
1,1-Dichloro-ethylene	$C_2H_2Cl_2$	98	5	1–16	86, 149
1,2-Dichloroethane	$C_2H_4Cl_2$	99	5–37	38–1350	86, 149, 154, 158
Chlorotrifluoro-methane	CF_3Cl	104	4	—	153
1,3-Dichloropropane	$C_3H_6Cl_2$	112	—	—	89
Chlorobenzene	C_6H_5Cl	112	—	50–500	86
Chloroform	$CHCl_3$	118	8–21	13–225	86, 87, 90, 94, 114
Dichlorodifluoro-methane	CF_2Cl_2	120	300–331	300–2400	34, 149, 155, 159–161
1-Bromopropane	C_3H_7Br	128	—	—	92
Bromochloromethane	CH_2BrCl	128	2.3–3.1	—	162, 164
Trichloroethylene	C_2HCl_3	130	2–20	12–4000	86, 91, 92, 94, 135–149
1,1,1-Trichloro-ethane	$C_2H_3Cl_3$	132	80–156	140–10100	86, 90, 135, 149, 155, 160
1,1,2-Trichloro-ethane	$C_2H_3Cl_3$	132	5	4–45	86, 117
1-Bromobutane	C_4H_9Br	136	—	—	92
Trichlorofluoro-methane	$CFCl_3$	136	182–200	180–6000	34, 86, 155, 159, 160
Hexafluoroethane	C_2F_6	138	4	—	153
Methyl iodide	CH_3I	142	1–3	1–11	149, 152, 154, 155
1,2-Dichlorobenzene	$C_6H_4Cl_2$	146	—	1–50	56, 85–87, 90
1,3-Dichlorobenzene	$C_6H_4Cl_2$	146	—	0.5–25	86
Bromotrifluoro-methane	$CBrF_3$	148	0.8	—	154, 164

Table 2.12. (*Continued*)

| Compound | Formula | Mol. wt. | Concentration, ppt | | Ref. |
			Background	In the urban air	
Tetrachloromethane	CCl_4	152	95–135	95–1400	86, 133, 135, 149, 159
Chloropentafluoro-ethane	C_2F_5Cl	154	4	—	153
Tetrachloro-ethylene	C_2Cl_4	164	15–89	53–3700	56, 85, 86, 91–93, 149
Chlorobromodi-fluoromethane	$CBrClF_2$	164	0.9–1.2	—	162, 164
1,1,1,2-Tetrachloro-ethane	$C_2H_2Cl_4$	166	—	1–16	86, 158
1,1,2,2-Tetrachloro-ethane	$C_2H_2Cl_4$	166	—	1–96	86
Dichlorotetra-fluoroethane	$C_2F_2Cl_2$	170	10–14	—	154, 165
Dibromomethane	CH_2Br_2	172	3–60	—	163, 164
Trichlorotri-fluoroethane	$C_2F_3Cl_3$	186	16–25	16–4200	86, 165–168
1,1-Dibromoethane	$C_2H_4Br_2$	186	3	2–200	86, 154, 164, 169
Hexachloroethane	C_2Cl_6	234	—	—	158
Tribromomethane	$CHBr_3$	250	2–46	—	163, 164
Hexachlorobutadiene	C_4Cl_6	258	—	—	158

2.3.1 Chloro- and Chlorofluorohydrocarbons

The background concentrations of halocarbons in the air transported from the oceans to the continents and those in the upper layers of the troposphere are listed in Table 2.13. The table shows that about 25% of the total amount of chloride contained in organic components is present in the atmosphere in the form of methyl chloride which is predominantly of biogenic origin and the other 75% fall to the share of other chlorine-contained compounds emitted from anthropogenic sources. Almost half of this quantity is contained in chloro-fluorohydrocarbons. This distribution was characteristic of the beginning of the 1970s.

In the 1970s, a drastic change in the background concentrations of a number of chlorofluorohydrocarbons was observed. This refers, first, to CFC's. For example, the $CFCl_3$ content increased from less than 80 ppb to about 170 ppt for the period from 1971 to 1979. The authors of ref. [170] on the basis of their own observations and the literature data evaluated the average annual increase in

Table 2.13. Background concentrations (ppt) of some halocarbons in the troposphere

Location of observations	Time of obs.	$CFCl_3$	CF_2Cl_2	CF_3CCl_3	CCl_4	CH_3Cl	$CHCl_3$	CH_3CCl_3	C_2HCl_3	C_2Cl_4	Ref.
USA, Washington state	IX 1974–II 1975	125±8	—	—	120±15	530±30	20±10	100±15	<5	20±10	149
California	XI–XII 1975	103.8	180.8	16.3	114.2	952.9	23.4	84.0	—	40.7	168
Pacific Ocean North-eastern part	III 1976	130±5	227±7	21±4	122±13	569±41	8.8±2.7	94.5±8.2	20±4	15.6±4.6	165
Shore of Japan	VII 1978	159	268	25	104	—	—	96	2	7	166
Eastern part: 40°N–0°	XII 1981	186	305	23	135	—	21	156	12	29	154
0°–40°S	XII 1981	172	282	21	128	—	11	116	<3	5	154
Antarctica, Palmer station, 64°46'S	I 1983	189.8	314.4	—	148.9	—	—	118.6	—	—	170
	I 1984	199.6	329.0	—	150.2	—	—	—	—	—	159
The Sea of Okhotsk, 44°58'N	IX 1985	201	295	—	124	—	28	—	—	—	*

*Author's data

$CFCl_3$ concentrations from 1971 to 1974 to be 21%. Pack D.H. et al. [171] analyzed the information on the content of this substance in the atmosphere obtained for five years (1970–1975) and have reported that it increased at a rate of 13.2% per year in the Northern Hemisphere and 14 to 6% in the Southern Hemisphere. The monitoring of $CFCl_3$ for two years (1976–1977) by Australian researchers allowed them to draw the conclusion that the average annual trend is 19%.

The concentrations of CF_2Cl_2 increased in the troposphere at approximately the same rates in the 1970s. However, at the end of the past and at the beginning of the present decade the rate of increase in the CFC content became much less pronounced. According to the data in ref. [54], the average concentrations of $CFCl_3$ and CF_2Cl_2 in the moderate latitudes of the Northern Hemisphere increased from November 1979 to December 1981 by 15 ± 3 ppt and 26 ± 5 ppt, respectively, i.e. by 8 to 9% per year. The rate of increase in the content of these CFC's dropped by about one and a half (8.2 and 19.4 ppt per year) as was detected from August 1980 to February 1982 at Point Barrow (Alaska) [155]. It was reported for the first time in ref. [155] that the concentrations of CHF_2Cl and CH_3CCl_3 exhibit a considerable trend (11.9 and 7.9% per year, respectively). The increase in CCl_4 concentrations for the first half of the 1970s was 10.5% and 2.2% in the Northern and Southern Hemispheres, respectively [171].

The tendency to lower rates of increase in the concentrations of halocarbons was maitained during the entire first half of the 1980s. The results of measurements of concentrations at the Palmer station from February 1982 to December 1985 showed the unusual increase of 5.87% for $CFCl_3$, 5.45% for CF_2Cl_2, 5.31% for CH_3CCl_3 and 1.29% for CCl_4 [159]. The applications of aviation and research ships and the establishment of a network of control stations from Alaska to the South Pole has made it possible to detect the existence of a latitudinal concentration gradient of halocarbons. In the period from 1978 to 1981 it was 8 to 9% for $CFCl_3$ and CF_2Cl_2, 5% for CCl_4 and 37% for CH_3CCl_3 [172]. In the following years, a certain decrease in the latitudinal gradient was observed, which presumably reflects a certain decrease in the scale of anthropogenic emissions, in particular in the Northern Hemisphere.

The seasonal variations in concentrations very characteristic of light hydrocarbons are observed only for halocarbons containing hydrogen atoms or double bonds (for example, for CH_3CCl_3 and C_2Cl_4). Minimum amounts of these components are present in the atmosphere in summer time, the amplitudes of fluctuations being more pronounced in the Northern Hemisphere [173].

Numerous experiments on the determination of distribution of halocarbons throughout the atmosphere [34, 156, 164, 174–177] have shown that in the troposphere the concentrations of most of them only slightly depend on altitude. The concentrations decrease on the whole in the tropopause and in the lower layers of the stratosphere. This decrease is particularly appreciable for 1,1,1-trichloroethane, methyl chloride and tri- and tetrachloroethylenes [64]. However, the change in the content of all components is not monotonic: the concentration profile in the upper troposphere and the lower layers of the

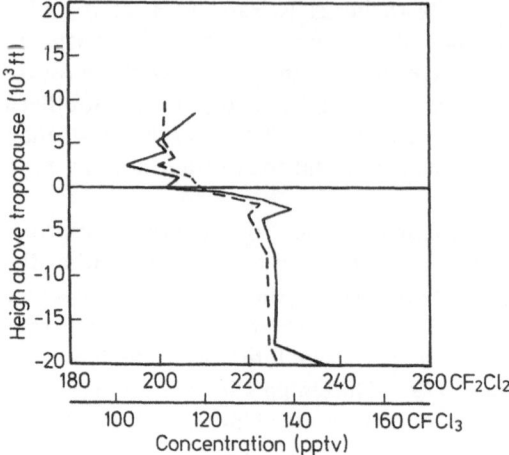

Fig. 2.9. Concentration profiles of CF_2Cl_2 (—) and $CFCl_3$(- - -) according to the data of analyzes carried out on March 12, 1976 [165]

stratosphere is much more complex as is shown in Fig. 2.9. The figure shows that in the layer of the troposphere about 1.5 km thick adjoining the tropopause, marked variations in the CFC content are observed. Above the tropopause, the concentration profile is also represented by a broken line reflecting the alternating decrease and increase in the amount of halocarbons.

The reason for a sharp change in the gradient between the neighboring layers is the instability of the lower layers of the stratosphere and the change in the height of the tropopause. In this transition layer, the streams of the tropospheric air enter the stratosphere and when this intrusion occurs, an increase in the content of halocarbons (as well as in that of other components emitted from the Earth's surface and reaching high altitudes) is observed. The decrease in

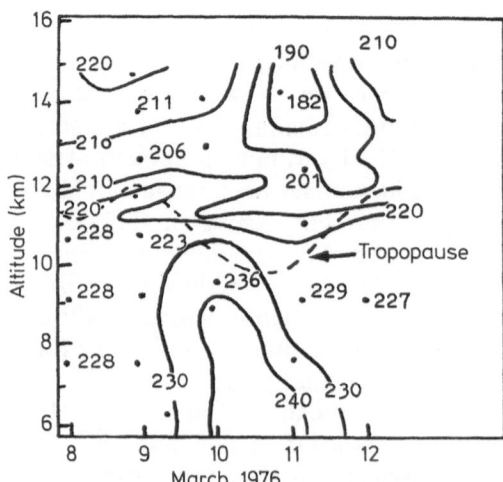

Fig. 2.10. Isopletes of CF_2Cl_2 concentrations in the period from 8 to 12 March 1978 above western regions of the U.S.A. [165]

concentration below the tropopause results from the penetration of the stratospheric air into the troposphere.

The concentration profile in these atmospheric layers probably seldom remains constant even for one day. The changes taking place from day to day are illustrated in Fig. 2.10 which shows the lines of equal concentrations of CF_2Cl_2 for several days in the mid-latitudes of the Northern Hemisphere over the USA. The figure shows that four days out of five the air layer at an altitude of 12–13 km with a relatively low CFC content was covered with another layer characterized by higher concentrations.

Hence, the distribution of minor components in the upper layers of the troposphere and higher than the tropopause is controlled by meteorological factors.

The exchange between the troposphere and the stratosphere proceeds differently in different latitudes. The most active intrusion of the tropospheric air into the stratosphere is evidently observed near the subpolar jet flow [64]. Another location of active intrusion of "parcels" of the tropospheric air is the intratropical convergence zone. The researchers [176] have detected the existence of a latitudinal gradient of the concentration of minor components in the lower stratosphere: when moving along the meridian from the intratropical convergence zone (7 °N) in the pole direction to 79 °N, a considerable decrease in CFC concentration has been observed. The increased content in $CFCl_3$, CF_2Cl_2 and methane over the tropics indicates the existence of a strong ascending flow intruding into the lower stratosphere through the tropic tropopause. The intruding masses of air are involved in meridional transport which may be regarded as a continuation of the trade wind Hadley cell in the stratosphere.

In the middle layers of the troposphere, most halocarbons are distributed relatively regularly, whereas in the nearground air layer over the continents, an increase in concentration is sometimes observed. An example of this situation is shown in Fig. 2.9. The study of the trajectory of motion of air masses has shown that in the period of sample collection the air was emitted from a densely populated areas [165].

The concentration level of halocarbons in the urban atmosphere varies considerably but usually greatly exceeds the background level. Comparison of figures listed in Tables 2.12 and 2.13 shows that most often the concentrations in the urban air greatly exceed the background concentrations.

2.3.2 Chlorine Pesticides

The problem of increasing the production of food-stuff and the productivity of labor in agriculture has led to the necessity for chemical plant protection. At present organic pesticides, substances intended for the extermination of weeds, pests and diseases of cultural plants and forests, are extensively used throughout the whole world.

Organic chlorine pesticides are a very effective means for plant protection. The most well-known among them are the compounds of the DDT type, isomer hexachlorocyclohexanes (HCH), derivatives of 2,4-dichlorophenol (2,4-D) and chlorinated polycyclic hydrocarbons (eldrin, dieldrin, endrin, chlordane, etc.)

p,p'-DDT   2,4-D-acetic acid   Dieldrin

The extremely high stability of these compounds in soil and water is the reason for their accumulation in various objects of the environment and the unfavorable effect on man and the animal world because they are very toxic not only for insects but also for vertebrates. For most of them, the halflife period in soil exceeds 1.5 years and for DDT and dieldrin it is 15–20 years. Herbicides and defoliants of the 2,4-D series are extremely dangerous because they are prepared by the procedure often resulting in the pollution of the final products by the admixtures of polychlorinated dibenzodioxines. These compounds are strong mutagenic and teratogenic agents.

Organic chlorine pesticides have a tendency to be accumulated in individual organs and fat tissues and pass along the trophic chain from one organism to another. As a result, the organism situated at the end of the trophic chain absorbs pesticides and often their more toxic metabolites in amounts exceeding the safe level. Since these toxins mainly affect the reproductive function, as a result of the wide application of pesticides some biological species have been almost exterminated.

Up to the 1980s, the first place among pesticides with respect to the scale of production and use in agriculture was occupied by DDT and α-hexa-chlorocyclohexane (α-HCH, lindane). Consequently, at present all the objects in the environment are everywhere polluted with residual amounts of these pesticides. The situation can be characterized by the fact that over the past 30 years about 3000 t of DDT has accumulated even in the snow cover of the Antarctica. The extreme danger of DDT to living organisms had been recognized, and in 1970s and the beginning of the 1980s in many industrial countries its use was drastically limited or even prohibited. However, the world production of DDT did not appreciably decrease because it increased in the developing countries of Latin America, Africa and Asia. According to the evaluations of specialists, at present the level of DDT production attains 0.1 $\times 10^6 \text{t y}^{-1}$. The production of α-HCH is even higher.

One of the principal paths for the spreading of organic chlorine pesticides is their migration in the atmosphere. The proofs of this transport have been repeatedly obtained in the analysis of air at the locations remote from the regions of pesticide application, e.g. on the atolls of the Pacific Ocean and in the Antarctica. In the air masses over the oceans, such pesticides as DDT, chlordane, dieldrin, γ-HCH, α-HCH and other hexachlorocyclohexane isomers have also been detected.

Table 2.14. Concentrations (ng m^{-3}) of some organochlorine pesticides in the air of background areas [178]

Location of observation	Time of observation	α-HCH	γ-HCH	p,p'-DDT	ΣDDT
Pacific Ocean	1977–1978	0.37–1.7	0.17–0.18	—	0.025–0.2
Indian Ocean	1980–1981	—	—	1.0–1.5	—
Persian Gulf	1976–1977	0.18–0.40	0.04–0.2	—	0.16–0.21
Red and Arabian Sea	1976–1977	0.03–1.8	0.02–0.67	—	0.005–0.58
Antarctica:					
Sabrian Shore	1981		0.12	—	0.24
Ballena Island	1981		0.17	—	0.19
Syowa Station	1982		0.049	—	0.020
Amundsen Sea	1982		0.044	—	0.022
USSR, biospheric reservations:					
Berezina (Byelorussia)	1982–1984	$\dfrac{0.01-2.7^a}{0.6}$	$\dfrac{0.01-1.5}{0.4}$	$\dfrac{0.03-1.5}{0.3}$	$\dfrac{0.06-2.3}{0.4}$
Prioksko-terrasny (Moscow district)	1984	$\dfrac{0.12-2.5}{0.8}$	$\dfrac{0.05-1.5}{0.5}$	$\dfrac{0.13-1.6}{0.7}$	$\dfrac{0.16-2.1}{0.9}$
Caucasian (Krasnodar Territory)	1984	$\dfrac{0.05-1.4}{0.4}$	$\dfrac{0.09-1.4}{0.4}$	$\dfrac{0.06-1.8}{0.7}$	$\dfrac{0.3-3.5}{1.7}$
Repetek (Turkmen SSR)	1982–1984	$\dfrac{0.01-1.9}{0.4}$	$\dfrac{0.01-1.2}{0.2}$	$\dfrac{0.03-0.8}{0.1}$	$\dfrac{0.06-1.4}{0.3}$
Chatkal (Uzbek SSR, Tien Shan)	1984	$\dfrac{0.1-33.5}{7.0}$	$\dfrac{0.01-4.1}{2.0}$	$\dfrac{0.03-0.8}{0.2}$	$\dfrac{0.06-1.2}{0.3}$
Sary-Chelek (Kirgiz SSR, Tien Shan)	1984	$\dfrac{0.05-14.1}{2.7}$	$\dfrac{0.05-4.1}{0.7}$	$\dfrac{0.03-2.7}{0.2}$	$\dfrac{0.06-3.3}{0.3}$
Borovoe (Kazakh SSR)	1983, July	$\dfrac{0.08-0.9}{0.4}$	$\dfrac{0.04-0.6}{0.3}$	$\dfrac{0.03-0.5}{0.2}$	$\dfrac{0.06-0.6}{0.3}$
Czekhoslovakia, Koshetize	1980, July	$\dfrac{0.03-0.34}{0.14}$	$\dfrac{0.04-0.36}{0.19}$	$\dfrac{0.05-1.1}{0.44}$	$\dfrac{0.06-1.9}{0.69}$
Bulgaria, Ropotamo	1982, May	$\dfrac{0.49-4.4}{1.6}$	$\dfrac{0.56-5.8}{1.6}$	$\dfrac{0.23-2.3}{0.78}$	$\dfrac{0.60-2.86}{1.3}$
DDR, Neuglobzow	1984	$\dfrac{0.28-1.10}{0.64}$	$\dfrac{0.70-2.2}{1.3}$	$\dfrac{0.03-0.53}{0.33}$	$\dfrac{0.12-1.15}{0.75}$
Baltic Sea	1983, April	$\dfrac{0.02-0.52}{0.16}$	$\dfrac{0.03-0.45}{0.14}$	$\dfrac{0.03-0.90}{0.35}$	$\dfrac{0.09-1.51}{0.55}$

[a]The numerator gives concentration limits and the denominator shows average concentration

Table 2.14 lists the results of observations of the content of some pesticides in the overland air of various regions taken from review [178].

Apart from the literature data [179–182], we can report the results of our own measurements carried out in the biospheric reservations on the territory of the USSR and performed during joint expeditions in the background regions of some countries of Eastern Europe. The table gives total concentrations (in the vapor and aerosol phases) of DDT and HCH isomers. A series of observations carried out for many years has suggested that isomeric HCH's are mainly present in air in the form of vapor. In the case of DDT, the contribution of the vapor phase is also very great (not less than 50%).

It can be seen from Table 2.14 that pesticide concentrations in the air of hinterland regions of the USSR are not high. It may be noted that in the Europen part of the country the residual DDT content is two or three times greater than that over the Central Asiatic republics. In contrast, the concentrations of α-, γ-HCH's are much higher over the Asiatic territory of the USSR. These relatively higher levels of HCH's are probably due to its wide application instead of the prohibited DDT. The maximum HCH's concentrations are usually observed in the spring months.

2.4 Organo-Element Compounds

An organometallic compound, tetraethyl lead $Pb(C_2H_5)_4$, was used from the beginning of the 1920s as an antiknock additive to low octane-number gasoline. This substance is very toxic and, as medical investigations have shown, accumulates in brain tissues. It is emitted into the atmospheric air mainly with the exhaust gases of automobile transport and gasoline vapor.

Beginning in the 1960s, in all countries instead of tetraethyl lead a more highly volatile tetramethyl lead (TML) has been used. At present, in the air of some cities of Western Europe about 75% of organo-lead compounds are present in the form of TML [183].

The character of the principal source of organo-lead compounds determines their distribution in the atmospheric air of various regions. The highest concentrations attaining several micrograms per cublic meter have been detected near superhighways. In rural areas without heavy traffic, they are lower by about three orders of magnitude [184, 185]. In cities (Table 2.15), the greatest amounts of lead compounds are usually observed in the air of the business centers and near the parking lots and gas-filling stations. Until recently it was believed that lead is present in the urban atmosphere only in the form of inorganic salts contained in aerosols. The figures given in Table 2.15 show that the contribution of volatile organo-lead compounds is appreciable. The toxicity of organic compounds is much higher than that of inorganic substance and therefore the control of the content of the former is a very important problem.

In recent years it has been established that other forms of organo-lead compounds, tri- and dialkyl lead, are also always present in the atmosphere. The

Table 2.15. Concentrations $(ng\,m^{-3})$ of lead compounds in the urban atmosphere

| City | Lead concentration in: | | References |
	organic substances	inorganic substances	
Frankfurt on Main:			183
Business center	45 ± 34	690 ± 360	
Blocks of dwelling houses	24 ± 29	370 ± 510	
Parking lots	678 ± 253	3270 ± 500	
Antwerp:			189
Business center	76–262	311–1000	
Blocks of dwelling houses	8–20	134–309	
Filling station	192–213	597–679	
London	40–110	3200–8800	185
Lancaster	69	2560	186

concentrations of volatile trimethyl and triethyl lead in the urban air are approximately 10% of the corresponding tetraalkyl compounds. It is assumed that tri- and dialkyl substitute compounds are formed from tetraalkyl substituted compounds as a result of photochemical processes in the atmosphere [187].

Among other organometallics, attention is drawn to very toxic compounds containing mercury [188, 189]. They are methyl mercury, CH_3HgH, detected in motor vehicle exhaust gases, the water-soluble form of CH_3HgOH, methyl mercuric chloride, CH_3HgCl, and dimethyl mercury, CH_3HgCH_3. The latter two components are present in air at the concentrations of 0.004–$0.04\,ng\,m^{-3}$. Their formation is explained by microbiologic processes occurring during the treatment of sewage and industrial waste.

Microorganisms and microscopic algae are evidently responsible for the formation and emission into the atmosphere of a number of other organo-element compounds. For example, mono-, di- and tri-methyl arsenates have been detected in the water of estuaries of some English rivers at concentrations attaining $1\,\mu g\,l^{-1}$ [190]. Special investigations have shown that the methylation of inorganic arsenites and arsenates is carried out by phytoplankton [191]. The same source is probably responsible for the appearance in the air of organic selenium compounds $(CH_3)_2Se$ and $(CH_3)_2Se_2$ the concentrations of which vary from less than 0.15 to $2.4\,ng\,m^{-3}$ [192a].

2.5 Organic Compounds Present in Aerosols

Most organic compounds are not only present in the Earth's atmosphere in the form of vapor but also as aerosols i.e., solid or liquid particles varying in size from less than 0.01 μm to a few hundredths of a millimeter.

The total amount of aerosols emmited by various sources is considered to be about $2600\,Tg\,y^{-1}$. According to the opinion of many researchers, aerosols contain only a small part of the total amount of orgnic substances present in the atmosphere. This opinion is based on the results of numerous experiments on the determination of the chemical composition of aerosols, which indicate that their organic substance content is relatively low. However, one cannot be quite sure that the present analytical procedure adequately reflects the true composition of these particles.

The quantity of organic substances is determined by various methods (most often, by chromatographic methods) after the aerosols have been collected on filters. It may be assumed that a considerable part of the volatile components is lost during sample collection because they pass into the gas phase. This is particularly true for compounds present in liquid particles, the fraction of which is often large. For example, Cadle R.D. et al. [192b] reported that liquid drops predominate in an aerosol collected in Los Angeles. Goetz A. et al. [192c] reported that the number and size of solid particles collected in west of the USA decreased upon repeated investigation carried out several days after the collection. This decrease is probably due to the evaporation of volatile organic compounds. Hence, in our opinion, the evaluations of global emission of organic compounds in aerosols (about $56\,Tg\,y^{-1}$) [193] may be greatly underestimated.

The study of these compounds is of great interest for elucidating important problems of atmospheric chemistry. First, aerosols are optically active components profoundly affecting the radiation balance of the atmosphere. Their ability to absorb, reflect and scatter sunlight is related to a number of physicochemical properties which depend on their chemical composition. Second, aerosols of the submicron size (Aitken condensation nuclei) play a decisive role in the condensation of water vapor and the formation of the cloud cover of the planet. Third, the conversion of gaseous components into solid particles with subsequent coagulation and sedimentation is one of the principal mechanisms for the purification of the atmosphere. Finally, the study of aerosols is very important because they contain the predominant amount of many very toxic organic substances, in particular, polyaromatic hydrocarbons (PAH).

The ratio of substance concentration in the gas phase to that in the condensed phase depends mainly on the vapor pressure at the ambient temperature. According to Junge's suggestion, aerosols should contain over half of the total amount of a compound if the pressure of the saturated vapor for this compounds is lower than $10^{-7}-10^{-8}$ mm Hg. This value is characteristic of normal paraffins with the number of carbon atoms in the molecule more than 25 at a temperature of 293 K. It has been established experimentally that in summer time n-alkanes with the number of carbon atoms less than 18 are virtually not retained by aerosol filters, and for higher homologues the ratio of aerosol content in the condensed and the vapor phases in the summer time varies from 0.053 for n-$C_{19}H_{40}$ to 2.8 for n-$C_{31}H_{64}$ [194]. The authors of ref. [194] have reported that the amount of n-alkanes heavier than eicosane ($C_{20}H_{42}$) in the form of vapor is negligible even under the conditions of mild winter usual for the Atlantic shore of Europe. In summer time, the gas phase virtually did not contain paraffin C_{24}–C_{25} hydrocarbons and heavier homologues. However, German researchers analyzing the air over the northern Atlantic Ocean and the southern Indian Ocean continuously detected much larger amounts of C_{20}–C_{28} n-alkanes in the gas phase than in aerosols [67, 195].

The material of aerosol particles is heterogeneous: they contain organic and inorganic compounds. Their ratio often depends on the origin of aerosols. The particles emitted into the atmosphere as a result of weathering processes contain silicate materials and oxides forming rocks and the mineral basis of soils. These aerosols contain a considerable amount of organic substances which are humus components. Oceanic aerosols formed by dispersion of sea water contain various salts. The finest drops formed when bubbles on wave crests are destroyed and lose water by evaporation, and the remaining particles consist of salt and low volatile organic substances present in sea water.

In addition to dispersion processes, the condensation phenomenon also plays an important role in the formation of aerosols. The condensation of carbon during combustion under the condition of insufficient oxygen supply leading to the isolation of finest soot particles may serve as an example.

Small particles constantly present in the atmosphere serve as nuclei on the basis of which (probably also as a result of condensation processes) larger particles begin to grow. Strictly speaking, the mechanisms for the growth of aerosol particles are unknown. The processes of conversion of gaseous organic components playing a very important role in chemical reactions occurring in the urban atmosphere and leading to the formation of smog are particularly poorly investigated.

As already mentioned, aerosols are constantly present in the atmosphere but their content in air over different regions is different. The background concentration over the oceans of both hemispheres varies from 9 to $150 \mu g \, m^{-3}$; however, its ranges are usually more narrow: from 20 to $30 \mu g \, m^{-3}$ [67]. In arid zones and at sea shores it can increase to $500 \mu g \, m^{-3}$.

The content of aerosol particles in urban air is profoundly affected by their size, the level and type of industrial production and by the season. In the air of New England cities (USA) in the 1970s, this content varied from 20 to $330 \mu g \, m^{-3}$. Moreover, the highest concentrations were invariably observed in winter time [196, 197]. In the air of the city of Hamilton, a large metallurgic center in the Ontario province of Canada, it was 60–146 and in Toronto it varied from 60 to $137 \, \mu g \, m^{-3}$ [198].

A number of investigations carried out in recent years confirm the fact that organic compounds are one of the main components of the atmospheric aerosols. The sum of organic (noncarbonate) carbon is 3–25 % and can sometimes exceed 60% of the total mass of particles. Comparison of aerosols collected over the seas and the continents shows that in the latter case the total amount of organic carbon increases greatly owing to the presence of partly oxidized material (resin), humus components and fragments of vegetable tissues.

Compositions are usually compared according to the results of determination

Table 2.16. Organic carbon content in background aerosol [193]

Sample location	Organic carbon concentration, $\mu g\,m^{-3}$	
	mean	range
North Atlantic Ocean	0.76 ± 0.42	0.33–1.6
West Ireland	0.57 ± 0.29	—
The Bermuda Islands	0.37 ± 0.23	0.15–0.78
— " —	0.29 ± 0.09	0.15–0.47
The Hawaiian Islands	0.39 ± 0.03	0.36–0.43
The Pacific Ocean (10 °N)	0.49 ± 0.26	0.22–0.74
The Pacific Ocean (10 °S)	0.32 ± 0.18	0.07–0.53
Samoa	0.22 ± 0.09	0.13–0.41
Tasmania	0.53	—

of the content of organic substances that can be extracted by various solvents (benzene, methylene chloride, diethyl ether and methanol). The summary given in Table 2.16 shows that average concentrations of organic carbon in the background aerosol vary over a relatively narrow range from 0.2 to 0.9 $\mu g\,m^{-3}$. The data presented by various investigators do not show any considerable differences in organic carbon content in the background aerosol of the Northern and Southern Hemispheres.

The average concentrations of organic compounds in aerosols from non-urban areas over the continents of the Northern Hemisphere are 2–3 $\mu g\,m^{-3}$. Thus, according to the results of numerous analyses carried out in the 1960s–1980s in various parts of the USA, their average concentration was 2.2 ± 1.0 $\mu g\,m^{-3}$ with a scatter of 0.6–5.0 $\mu g\,m^{-3}$ [193, 199–203].

At present, our information on the altitude distribution of aerosols is very sparse. However, the available data suggest that organic compounds in aerosols are present throughout the atmosphere. In the 1970s, some authors carried out investigations in the mountain regions of Europe and America. Two samples were collected at a locations in the Swiss Alps and at another in the Bolivian Andes at the altitudes of 3600 and 5200 m above sea level [194, 204]. In the former case the average concentration of organic carbon was 1.1 $\mu g\,m^{-3}$. The total composition of aerosols in the Andes was not reported but the authors indicated that the quantity of C_{20}–C_{33} n-paraffins and PAH in this aerosol was 10–40 times less than in rural areas of Belgium and Holland which are strongly influenced by urban areas.

As was to be expected, the quantity of organic hydrocarbon in the aerosol emitted by urban air exceeds many times the background levels. The results of a number of investigations carried out in some cities of America and Europe are listed in Table 2.17.

The problem of distribution of atmospheric aerosols according to the size of

Table 2.17. Organic carbon content in urban aerosols

City	Concentration, $\mu g\,m^{-3}$ Average	Range	References
USA:			
Los Angeles	—	4.7 ± 40.4	196
— " —	—	7–46	205
Pasadena	26 ± 7	22–34	206
Houston	—	1.2–21.8	196
Cities of New England	—	1.8–30.4	196
Canada:			
Hamilton	—	4.3–23.8	198
Toronto	—	6.0–12.8	198
Mexico, Mexico City	7.2 ± 1.8	1.9–21.9	207
German Federal Republic, Mainz	33	—	206
England, London	12	9–14	208
Turkey, Ankara	63	56–69	208

their particles and the content of organic components in their various fractions is very important from the geophysical and medicohygienic aspects. The particle size determines the time of residence of organic compounds in an aerosol in the atmosphere. The particles the diameter of which exceeds several μm are relatively rapidly removed from the atmosphere as a result of sedimentation and impaction (removal as a result of collision with various obstacles: house walls, trees, rain drops, etc.). The finest particles less than 0.01 μm in size are also shortlived because of their tendency to coagulate. However, the aerosol particles with sizes varying from approximately 0.04 to 1.0 μm are virtually not involved in the coagulation processes and do not undergo rapid sedimentation. Hence, their residence time in the atmosphere is relatively longer: several tens of hours.

On the other hand, aerosol size determines their ability to penetrate into the respiratory system and being retained in it. It has been established that particles about 5 μm in diameter or larger are retained in the upper part of the respiratory tract of man, aerosols of smaller diameter penetrate into the bronchi and those 1 μm or less in diameter penetrate directly into the alveoli.

Table 2.18 demonstrates the differences in the chemical composition of fine and large particles. It gives the results of analysis of two fractions of the background aerosols collected in the west of the USA [203]. It can be seen that the main element in the fine fraction is sulfur. Ion exchange chromatography and protentiometric titration have been used to show that 61% of the fine particles is made up of ammonium bisulfate, NH_4HSO_4. In large particles, the elements contained in the Earth's crust, silicium, aluminium, iron and calcium, predominate. Organic carbon represents 11.4% of the total mass of aerosols and is mainly contained (65%) in the fine fraction.

Hoffmann E.J. and Duce R.A. have reported the same character of organic

Table 2.18. Mass and elemental composition of two fraction of aerosol particles from the Great Mountains National Park (USA) [203]

Element	Concentration, $\mu g\,m^{-3}$	
	$\leqslant 2.4\,\mu m$	$2.4-20\,\mu m$
$C_{organic}$	2200	1200
$C_{elemental}$	1100	< 100
Al	20	195
Si	38	580
S	3744	204
Cl	< 10	7
K	40	108
Ca	16	322
Ti	< 2	18
V	—	2
Fe	28	118
Ni	1	1
Cu	3	< 5
Zn	9	< 4
As	2.2	< 1
Se	1.4	0.2
Br	18	5
Pb	126	14

carbon distribution in oceanic aerosols collected in the region of the Bermuda and Hawaiian Islands and Samoa: over 80% of organic carbon has been detected in particles less than 1 μm in diameter. According to the reports of various authors, from 75 to 95% of organic compounds contained in the urban aerosols form part of the particles smaller than 3 μm [193]. Hence, the most longlived and the most harmful (because of their ability to deeply penetrate into the respiratory tract of man) particles are enriched with organic compounds.

The individual composition of organic aerosol components is very complex: they contain more than 500 compounds of various classes, such as aliphatic, alicyclic and polynuclear aromatic hydrocarbons, many hydrocarbon derivatives: aldehydes and ketones, alcohols, acids and esters, chlorine-, nitrogen- and sulfur-containing compounds and their polyfunctional derivatives.

The organic components of aerosols are usually separated by column chromatography into neutral, acidic and basic fractions. The first fraction, in turn, is separated into three parts: paraffin hydrocarbons, PAH and polar compounds. The authors of refs. [67, 195, 204] have shown that the group composition is relatively constant if the short fluctuations in time and space are neglected. In the oceanic, continental and urban background aerosols, the neutral, acidic and basic fractions represent 50–70%, 20–40% and 1–20%, respectively, of the total amount of compounds extracted with diethyl ether.

2.5.1 Alkanes

According to the reports of the authors of refs. [67, 195] who studied the group composition of the background oceanic aerosols of both hemispheres, about 50% of the "neutral" fraction is made up of aliphatic compounds. In this group, the alkanes with a straight chain of carbon atoms have been investigated in greatest detail.

The total concentration of n-alkanes in the air over the Atlantic Ocean was 2.7–7.8 ng m^{-3} and that over the Indian Ocean was 10 ng m^{-3} [195]. In 1971–1974, Simoneit B.R.T. detected $C_{16}H_{34}$–$C_{35}H_{72}$ n-alkanes in approximately the same quantities the sum of which did not exceed 10 ng m^{-3} in the dust particles collected in the Atlantic Ocean off western Africa [209].

The individual composition of n-alkanes isolated from background aerosols is characterized by the following features:

1. The concentrations increase from about 0.1 ng m^{-3} for hydrocarbons with a small number of carbon atoms to about 0.5 ng m^{-3} for the highest homologues.
2. The concentration increase is not monotonic: among the homologues with the number of hydrogen atoms 2n − 1, 2n and 2n + 1, odd-numbered homologues usually have higher concentrations.

For the qualitative evaluation of the relative contribution of odd- and even-numbered components, the parameter CPI (Carbon Preference Index) is used. It is calculated from the equation

$$ \text{CPI} = 0.5 \left[\frac{\sum\limits_{i}^{z} C_0}{\sum\limits_{i-1}^{z-1} C_e} + \frac{\sum\limits_{i}^{z} C_0}{\sum\limits_{i+1}^{z+1} C_e} \right], $$

where C_0 are the concentrations of homologues with the odd number of carbon atoms from i to z and C_e are the concentrations of homologues with the even number of carbon atoms. CPI is an important criterion which often makes it possible to determine the sources from which organic substances are emitted into the atmosphere (see chap. 3).

In the oceanic air having no contact with land for several days, n-alkanes contained in aerosols usually had the average CPI equal to 0.9, whereas the air transported from the continents brings hydrocarbons with a higher CPI value. For example, the maximum concentrations of n-alkanes in the air masses emitted from the African continent have been detected for C_{27}, C_{29} and C_{31} homologues, and the value of CPI ranged from 3 to 10 (Fig. 2.11).

The content of organic components (including n-alkanes) in aerosols collected

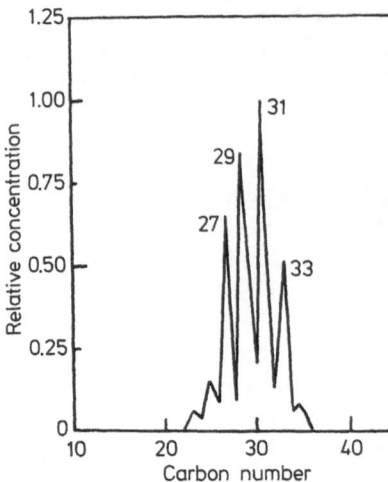

Fig. 2.11. Distribution of *n*-alkanes in atmospheric aerosol off the western shores of Africa [209]

in the air of inland seas is usually higher than that for oceanic aerosols. According to the data in ref. [210], the total concentrations of C_{15}–C_{32} *n*-alkanes in aerosols from the southern part of the Baltic sea in August–September 1983 ranged from 25 to 1000 ng m^{-3}. These high concentrations are due to the vicinity of powerful continental sources of organic atmospheric pollutants. The authors of ref. [211] reported a higher content of *n*-alkanes in aerosols collected in September–October 1983 over the western part of the Mediterranean Sea in those cases when the trajectories of movement of air masses intersected the countries of Northern Europe.

Table 2.19 shows the content and distribution of *n*-alkanes in aerosols collected over continents. It lists the individual and overall concentrations of *n*-alkanes both in the urban aerosols and in the particles from the agricultural areas more or less remote from large cities. Attention should be drawn to the existence of seasonal variations in concentrations: in winter time they are much higher than in summer, and are most pronounced for the urban atmosphere. The lowest concentrations have been observed in the air at the shore of the North Sea (Petten) and in mountains. In a rural area (Botrange), the concentration level of normal paraffins was found to be relatively high compared to that in mountains and over the seas, which indicates that urban areas profoundly affect this level.

Little information is available about the distribution of *n*-alkanes in aerosol particles with different dimensions. According to the data in ref. [212], in winter time the air of cities and their suburbs contains 80–95% of C_{22}–C_{32} *n*-alkanes in particles less than 3 μm in diameter, and the finest fraction (< 0.5 μm) contains 30–55% of these hydrocarbons. In winter it was observed that fine fractions were also enriched with C_{28}–C_{21} homologues but in summer they were distributed rather regularly in particles of all sizes (from less than 0.5 to 5 μm and larger).

Normal paraffins are not the only representatives of the class of saturated hydrocarbons present in aerosols. Apart from them, aerosol particles contain a

Table 2.19. Average concentrations (ng m^{-3}) of n-alkanes in aerosols from various locations of Belgium, Holland and Bolivia [194]

n-Alkanes	Urban air				Air of nonurban areas			Bolivian Andes	
	Antwerp		Ostend		Botrange[a]		Petten[b]	1	2
	Winter	Summer	Winter	Summer	Winter	Summer			
$C_{16}H_{34}$-$C_{19}H_{40}$	x[c]	x	x	x	x	x	0.57	—	0.20
$C_{20}H_{42}$	x	x	x	x	x	x	0.12	0.19	0.20
$C_{21}H_{44}$	1.94	x	x	x	x	0.47	0.06	0.19	0.40
$C_{22}H_{46}$	x	x	x	x	x	x	0.05	0.32	0.40
$C_{23}H_{48}$	5.20	x	2.15	x	x	x	0.16	0.43	0.65
$C_{24}H_{50}$	11.9	2.45	7.07	x	2.07	0.37	0.19	0.60	1.10
$C_{25}H_{52}$	17.71	4.31	12.29	0.41	5.29	2.76	0.74	0.50	1.10
$C_{26}H_{54}$	17.87	6.20	14.37	2.71	7.08	3.23	1.24	0.70	1.10
$C_{27}H_{56}$	21.01	9.12	17.17	3.28	7.55	3.99	1.52	0.40	0.70
$C_{28}H_{58}$	16.20	5.63	13.34	2.44	5.65	2.92	0.51	0.30	0.70
$C_{29}H_{60}$	23.21	16.83	8.41	5.43	7.55	5.95	4.56	0.50	1.00
$C_{30}H_{62}$	12.32	4.39	9.22	2.11	3.94	2.20	0.29	0.30	0.60
$C_{31}H_{64}$	20.87	17.98	14.51	5.60	5.61	5.42	3.90	0.20	0.40
$C_{32}H_{66}$	—	—	—	—	—	—	0.14	0.16	0.40
$C_{33}H_{68}$	—	—	—	—	—	—	0.73	0.16	0.31
Sum	148.32	66.91	98.52	21.98	44.74	27.31	14.78	4.45	8.86
CPI	1.54	2.78	1.24	2.03	1.47	2.12	4.61	0.76	0.99

[a] Rural area on the frontier between Belgium and German Federal Republic.
[b] Holland, shore of the North Sea.
[c] Below detection level.

multicomponent mixture of branched alkanes which are difficult to separate and also cycloalkanes [199–201, 213]. The relative content of n-alkanes, their isomers and cyclic hydrocarbons probably differs greatly in aerosols of different origins. According to a communication of Simoneit B.R.T. et al., in 1978–1979 in a sparsely populated area in the east of the USA, n-alkanes made up less than half (23–45%) of the hydrocarbon fraction, whereas they predominated in particles collected off the Atlantic shore of Africa.

2.5.2 Polynuclear Aromatic Hydrocarbons

Polynuclear aromatic hydrocarbons (PAH) are one of the most intensively studied of organic atmospheric components. This interest is due to the world-wide increase in lung cancer disease, especially noticeable in industrial countries. The researchers explain the increase in the amount of malignant diseases by the wide spreading in the environment of compounds of various classes having carcinogenic properties. Many PAH are some of the most active carcinogenic agents. The carcinogenic properties are also exhibited by benzopyrenes, dibenzo-pyrenes, benzo(b)fluoranthene, benzo(k)fluoranthene, benzo(a)anthracene, dibenzo(ac)anthracene, dibenzo(ah)anthracene and some other aromatic hydro-carbons with condensed rings. Moreover, it was found that many PAH and their metabolites display mutagenic activity.

In the last few years, owing to the development of highly efficient methods of gas chromatographic separation, it has been possible to determine the qualitative composition of the fraction of PAH in atmospheric aerosols. As a result of many detailed chromatographic-mass spectrometric investigations [211, 214–217], more than 150 PAH with molecular weights from 128 to 302 containing from two to seven aromatic rings have been identified in aerosols. They are represented by all three types of condensed compounds: linearly condensed (napthalene, anthracene, tetracene, etc.) angularly condensed (phenanthrene, chrysene, picene, etc.) and peri-condensed (pyrene, benzopyrene, coronene, etc.) compounds:

Anthracene   Pyrene   Chrysene

Apart from these principal PAH, large amounts of alkyl-substituted homologues and derivatives obtained by the hydrogenation of one or several rings have also been detected in the atmospheric aerosols. For example, in addition to chrysene, benzopyrene and benzofluoranthene, nine isomeric methylchrysenes, 11 meth-ylbenzopyrenes and 11 methylbenzofluoranthenes have been detected.

Just as higher paraffins, PAH are distributed between the gas and the condensed phases. According to ref. [194], the contribution of PAH with the molecular weight of 252 ($C_{20}H_{12}$) and higher to the gas phase is not large; for benzofluoranthenes it does not exceed 10% even in summer time. PAH with the

Table 2.20. Background concentrations of polynuclear aromatic hydrocarbons (ng m^{-3})

PAH	Point Barrow, Alaska 1979, (218) March	Point Barrow, Alaska 1979, (218) August	Petten, Netherlands (194)	Bolivian Andes (194)	USSR, 1984 (219)[a] Beresina biospheric reservation	USSR, 1984 (219)[a] Prioksko-Terrasnyi biospheric reservation	USSR, 1984 (219)[a] Borovaya station	USSR, 1984 (219)[a] Caucasian biospheric reservation
Phenantrene	0.12	0.017 }	0.16 }	0.055	0.02–0.56	0.01–0.33	0.01–0.61	0.10
Anthracene	0.01	0.01						
Fluoranthene	0.29	0.028	2.25	0.032	0.47–4.02	0.46–0.79	0.12	—
Pyrene	0.31	0.029	0.16	0.034	0.12–7.80	0.06–1.23	0.05–1.17	0.10–0.50
Chrysene	0.08	0.005 }	0.34 }	0.063	0.34–7.21	0.19–1.01	0.31–10.4	0.29–0.69
Benzo(a)anthracene	—			0.024	—	—	—	—
Methylpyrene	—		0.04	0.055	0.06–1.54	0.05–0.54	0.02–0.48	0.10–0.53
Benzofluoranthenes	—	*[b]	0.70	0.065	0.21–1.55	0.09–0.38	0.11–1.81	0.34–0.56
Perylene	0.01	*	0.02		0.01–0.40	0.01–0.10	0.001–0.06	0.003–0.03
Benzo(a)pyrene	0.03	0.01 }	0.46 }	0.065	0.09–1	0.04–0.73	0.04–0.43	0.04–0.34
Benzo(e)pyrene	0.21	*		—	—	—	—	—
Benzo(ghi)perylene	0.07	0.01	—	—	0.11–2.51	0.10–0.87	0.04–0.62	0.08–0.56
Coronene	0.02	0.048	—		0.03–0.20	0.02–0.11	0.02–0.10	0.02

[a] Concentration range in average monthly samples in biospheric reservations in 1984
[b] Below detectable level

molecular weight 223 (chrysene and benzoanthracene) in summer time are distributed between the two phases approximately equally. An even larger contribution to the gas phase should be observed for $C_{12}H_{12}-C_{17}H_{14}$ hydrocarbons. However, the analysis of air in the tropical latitudes over the Atlantic Ocean has shown an unexpectedly high PAH content in the gas phase. The authors of ref. [216] have reported that the total concentration of 11 PAH in the gas phase varied from 7 to 18 ngm^{-3}, whereas in aerosol particles it was only 0.1–0.2 ngm^{-3}.

PAH are detected in aerosols all over the world. Table 2.20 gives average concentrations of some PAH detected at the background stations in Alaska, in the mountains of South America and in various areas of Europe. It can be seen from the table that the content of these compounds in air depends on their distance from large cities and industrial areas. Special investigations have shown that PAH contained in aerosols collected in the arctic latitudes are emitted mainly from the areas situated at a distance of several thousand kilometers [218]. The microscopic analysis of aerosols collected in Alaska in summer time has made it possible to detect in them a considerable amount of pollen characteristic not of the vegetation of the Arctic tundra but of deciduous forests growing in the south-east of Canada and in the USA. Large distance migrations of PAH in aerosol particles have also been established by many other researchers [210–213, 216].

At present, extensive data on PAH content in the urban atmosphere are available [194, 198, 205, 207 215, 220–223]. In modern cities, the concentrations of these compounds vary over very wide ranges. They depend on the geographic situation, topography, the city size, the character of industrial production and the type of fuel used. Moreover, the concentrations of PAH undergo considerable seasonal variations. For example, the total concentrations of PAH recorded in the atmosphere of Antwerp and Ostend in summer time were 46 and 9 ngm^{-3},

Table 2.21. Distribution of PAH in various fractions of atmospheric aerosol in the city of Hamilton [198]

Hydrocarbon	Concentration, μg m^{-3}				
	< 1.1 μm	1.1–2.0 μm	2.0–3.3 μm	3.3–7.0 μm	> 7.0 μm
Benzo(ghi)perylene	4.19	4.18	3.43	1.76	1.62
Benzo(a)pyrene	0.72	0.89	1.26	0.46	0.41
Benzo(e)pyrene	0.98	0.46	0.55	0.17	0.18
Naphtho(1,2,3,4-def)-chrysene	0.91	0.48	0.34	0.11	0.11
Benzo(rst)pentaphene	0.12	0.10	0.07	0.06	0.05
Benzo(k)fluoranthene	0.63	0.27	0.31	0.13	0.10
Perylene	0.16	0.09	0.10	0.03	0.03
Dibenzo(def,mno)chrysene	0.02	0.06	0.01	0.01	0.006

respectively, and increased in winter to 405 and 225 ng m^{-3} [194]. In Canadian cities, Hamilton and Toronto, the content of PAH increased from 9.8 to 5.5 ng m^{-3} in summer and to 32.5 and 17.5 ng m^{-3} in winter, respectively [198]. The amplitude of seasonal variation is different in different cities. This is mainly due to different contributions of individual sources to the formation of the organic component of aerosol. These problems will be considered in greater detail in chap. 4.

Table 2.21 gives a typical picture of the distribution of eight PAH in various fractions of urban aerosols. It shows that 72–89% of the total quantity of PAH is contained in the finest particles less than 3.3 μm in size, which can deeply penetrate into the respiratory system of man.

2.5.3   Alcohols, Carboxylic Acids and Esters

Aliphatic and aromatic alcohols, carboxylic acids and their esters are constant components of atmospheric aerosols. Simoneit B.R.T. [209] have established that dust particles collected off the African shores contained C_{12}–C_{34} n-alcohols the total concentrations of which varied from 0.01 to 10 ng m^{-3}. The concentrations of C_{12}–C_{32} n-carboxylic acids in these samples were 0.001–30 ng m^{-3}. Just as for normal alkanes, the concentrations of higher alcohols and acids do not vary monotonically, but in this case the highest concentrations are observed for even-numbered homologues. Figure 2.12 shows a typical distribution pattern of fatty alcohols and carboxylic acids. It can be seen that n-$C_{28}H_{57}OH$ dominates among alcohols, and n-$C_{16}H_{32}O_2$ and n-$C_{26}H_{52}O_2$ homologues are the most abundant among acids.

In addition to these compounds, secondary C_{15}–C_{23} alcohols were detected as admixtures in the same aerosol. The highest concentrations among them

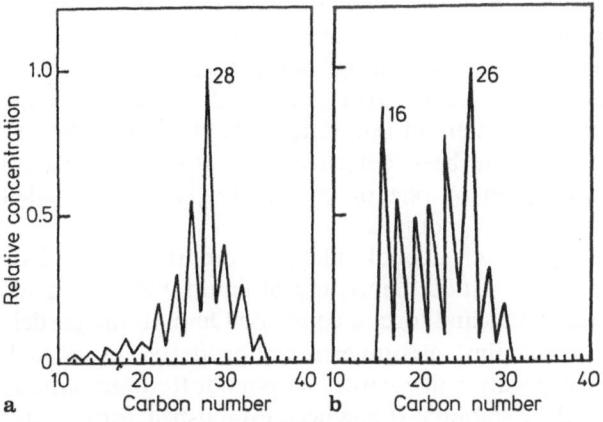

Fig. 2.12a, b. Distribution of alcohols (a) and carboxylic acids (b) in atmospheric aerosol off the western shores of Africa [209]

were observed for isononadecanol, $C_{19}H_{39}OH$, and mono-unsaturated C_{16}–C_{18} acids.

Much larger amounts of aliphatic alcohols and acids have been detected by Simoneit B. R. T. et al. in continental aerosols collected in a nonurban area in the east of California [199]. The total concentrations of normal alcohols varied from 198 to 524 ng m^{-3} and those of fatty acids ranged from 90 to 390 ng m^{-3}. In the same paper the figures characterizing the content of acids and alcohols in aerosols from Los Angeles air are given for comparison. They are 300 and 1500 ng m^{-3}, respectively and their sum makes up about 40% of the total amount of organic compounds in aerosols. Apart from monocarboxylic acids, considerable amounts of aliphatic C_3–C_{10} dicarboxylic acids have been found in the atmosphere of this city. Of 15 identified compounds, the greatest relative contribution is provided by succinic (22–25%), glutaric (17–23%), methylsuccinic (12–14%) and adipinic (12–13%) acids. In recent years, information has appeared that aerosols also contain other bifunctional compounds, e.g. ω-oxocarboxylic acids [224].

High percentage of oxygen-containing derivatives is characteritic of cities in the atmosphere of which photochemical processes are very pronounced. The relationship between the concentration level of these compounds and the photochemical activity has been reported by the authors of ref. [196] who have detected the increase in the water-soluble fraction of organic compounds forming a part of aerosols in the period of strong smog in the atmosphere of Los Angeles.

Seasonal variations in the content of the higher n-C_{16}–C_{28} monocarboxylic acids have been detected in the atmosphere of Antwerp [225]. The greatest total quantities attaining 200 ng m^{-3} were observed in winter. In this season, the individual concentrations varied from about 5 to 50 ng m^{-3}. Palmitinic ($C_{16}H_{32}O_2$) and stearic ($C_{18}H_{36}O_2$) acids were present in greatest amounts. These authors have also established that the principal of carboxylic acids is concentrated in particles less than 0.5 μm in size.

Among other oxygen-containing compounds detected in aerosols, one can name esters of aromatic and fatty acids. Aerosol samples from various areas of the Ontario province in Canada contained a large number of methyl esters of saturated and unsaturated C_{10}–C_{25} carboxylic acids with molecular weight varying from 186 to 410. Among esters of aromatic acids, dimethyl-, diethyl-, dibutyl- and dioctylphthalates have been detected [226]. Aromatic alcohols present in aerosols are represented by phenol, cresols, $C_{13}H_{12}O_2$ methylene diphenols [135] and many other compounds.

The highest carboxylic acids and alcohols play an important role in the chemistry of aerosols, in particular those consisting of liquid and hydrophilic particles. All of them are surfactants and form a condensed film on the particles surface. The composition of this film determines many properties of aerosols. First, this film prevents the evaporation of volatile compounds from the particles and thus profoundly affects their stability. It has been established, for example, that water drops covered with a monomolecular layer of C_{14}–C_{17} aliphatic alcohols evaporate hundreds of times more slowly than pure water drops [227].

On the other hand, the film of surfactants limits the absorption of inorganic components (CO_2, NO_2, SO_2, etc.). In addition to the above effects due to the presence of a monolayer of surfactants, it may be assumed that this film has a certain influence on the absorption and scattering of radiation in various spectral ranges.

3 Natural Sources of Organic Components of the Atmosphere

The study of sources of organic compounds, their power, localization in certain areas of the Earth's surface and their dynamics is very important for the general description of complex physico-chemical processes taking place in the atmosphere. The total content of various compounds is determined by the balance between the activity of sources and the processes of removal (sinks) from the atmosphere. The change in the concentration of some of them under the influence of anthropogenic factors can result in a whole series of interrelated perturbations of the entire structure of atmospheric processes or its individual elements. These perturbations are difficult to predict: their prediction requires the knowledge of sources and their power since the values of emission rates of organic compounds are some of the principal initial data for various mathematical simulations used for the solution of this problem.

Here it is appropriate to give an example illustrating possible consequences of the change in the activity of sources for some compounds. It has been mentioned above that in recent years an increase in global methane concentration at a rate of about $2\% \, y^{-1}$ has been observed. If this rate of increase is maintained, this will evidently lead in the near future (in 35 years) to the doubling of its amount in the troposphere. According to calculations carried out by using a one-dimensional radiative-convective model, this factor alone can lead by the mechanism of the greenhouse effect (methane exhibits a strong absorption band with a center at $7.66 \, \mu m$) to an increase in the temperature of the Earth's surface by 0.3–0.4 K. As already mentioned, the change in the average surface temperature exceeding 0.1 K is considered to be appreciable.

The other aspect of the problem is related to chemical transformations of methane in the upper troposphere and the stratosphere. It is assumed that the chemical reactions of methane with nitrogen oxides in the upper layers of the troposphere yield ozone which can subsequently diffuse into the overland air layers. Moreover, this ozone also contributes to changes in the temperature of the Earth's surface by participating in the absorption of UV radiation re-radiated by it in the wavelength range of 9–$10 \, \mu m$. The increase in methane concentration can profoundly affect the chemical processes occurring in the stratosphere because it should lead to the change in the content of hydroxyl radicals which are the main agents deactivating many components capable of destroying the ozone screen. Furthermore, methane itself serves as a kind of trap for chlorine atoms evolved in the photolysis of chlorofluorohydrocarbons and other chlorine-containing compounds and thus can perturb the catalytic cycle of ozone depletion (these processes are considered in greater detail in chap. 5). At present, it is difficult to estimate the overall effect caused by the increase in global methane concentration but it is evident that the study of its sources and changes in the dynamics of their functioning is very important. The same can be said about other organic components of the atmosphere.

The sources of atmospheric compounds are usually divided into anthropogenic and natural sources. The later with which this chapter is concerned can in turn be separated into biogenic and geogenic sources. The former are related to the functioning of the living matter of the planet (biota) in the modern period. Geogenic emit of organic substances formed as a result of abiogenic synthesis or accumulated in the bowels of the Earth owing to the activity of living organisms in ancient geological epochs and emitted into the atmosphere without the participation of man.

It should be emphasized that this classification is tentative. As present, many biogenic sources are directly or indirectly influenced by human activities and may be singled out as a separate group of quasi-natural sources. The increase in the sowing area, the melioration of excessively damp lands, the irrigation of arid zones and the development of artificial cultural landscapes result in the perturbation of the structure of natural biogeocenoses, the impoverishment of species of the flora and fauna and finally in a change in the emission rate and the composition of the emitted compounds.

The components emitted by all these sources belong to primary minor gaseous components of the atmosphere in contrast to the compounds formed directly in the atmosphere as a result of chemical transformations and are called secondary pollutants.

3.1 Biogenic Sources

First, certain general information about biogenic sources should be presented. It mainly concerns the distribution of the living matter of the planet among individual reservoirs. The total mass of the biota is estimated by various authors to be from 1.8×10^{12} to 2.4×10^{12} t based on dry weight. It is mainly concentrated on land. The predominant amount of the biomass (probably more than 90%) is made up by the autotrophic photosynthesizing organisms. Different climatic conditions result in the non-uniform distribution of vegetable cover on continents and its different productivities characterized by the annual increase in dry organic matter (Table 3.1).

The principal part of the phytomass of the planet (about 82%) is concentrated in forests, although their area occupies only 39% of the territory of continents. The forests of the tropical belt constitute half of the phytomass of forest, whereas the forests of the boreal belt (i.e. region located in the temperate latitudes of the

Table 3.1. Productivity of natural vegetable covers of continents

Continent	Surface area, $10^6\,km^2$	Total productivity, $10^9\,t\,y^{-1}$	%	Average productivity, $t\,km^{-2}$
South America	17.8	37.2	26.4	2090
Africa	30.1	31.0	22.0	1030
Australia (with Oceania and New Zealand)	8.9	7.6	5.4	860
Asia	43.1	38.3	27.1	880
Europe	10.5	8.9	6.3	850
North America	22.0	18.1	12.8	820
Total:	140	141.1	100	950

Northern Hemisphere) make up about one fifth of this mass. The other 30% are represented by the forests of transition zones (subtropical, subboreal zones, etc.).

Soil microorganisms rank second in volume as a reservoir of the living matter. Their total biomass is unknown but there are some data characterizing their content in soils.

According to the data of various authors, the live weight of soil bacteria is very substantial. Moreover, the greatest amount of the biomass is contained in soil under natural vegetable cover, whereas arable lands are poorer in microorganisms. It has been reported that in the soils of various geographic regions of the USSR and other European countries, the fresh weight of bacterial cells in the arable layer ranges from 0.6 to 9 t ha^{-1}. In the non-chernozem zone of the USSR, the biomass of bacteria in the 15 cm layer of forest soils attains 16.5 t ha^{-1}. The content of microscopic fungi in soils under natural vegetable cover is also very high. Their biomass in the 5–10 cm layer varies from about 0.2 t ha^{-1} in podzol soils under the pine forests to 10 t ha^{-1} in gray forest soil under the deciduous forests [229, 230]. Hence, the total biomass of bacteria and fungi attains considerable values: up to 20 t ha^{-1} in natural soils and up to 10 t ha^{-1} in cultivated soils. The smallest amounts of microorganisms are evidently contained in peat soils of subarctic regions and in sandy deserts where the total biomass is only a few tens of kilograms per ha [230].

Apart from bacteria and fungi, various microscopic algae are present in soils. Their content in arable lands ranges from several hundred to several thousand kilograms, and in soils under forests it attains 15 t ha^{-1}.

The animal population of the globe is much more diverse than the vegetable population. The number of species of animals exceeds that of plants by a factor of four but the total biomass of animals is about 2×10^9 t, i.e., is approximately 0.1% of that of higher plants.

Almost up to the end of the 1970s, very little attention was directed to natural sources of organic atmospheric components. Most investigators in the field of atmospheric sciences shared the opinion expressed as far back as 30 years ago by

Altshuller A. according to which a much smaller amount (both in quantity and in variety) of organic compounds is emitted from natural than from anthropogenic sources. As to the significance of the biogenic source, a decisive opinion has been expressed: "growing vegetation should not make any appreciable contribution to the natural hydrocarbon content of the atmosphere".

This concept was based on intuition deceived by the almost catastrophic state of the atmosphere of some large cities before measures were taken for its protection, rather than on specific experimental data. Unfortunately, the investigations of biologists who detected emission of organic compounds on a considerable scale by vegetation even in the first half of our century and advanced hypotheses on the role of these compounds in nature [79, 231–233a] did not attract the attention of specialists in atmospheric chemistry which had just begun to develop.

The emission of organic compounds into the environment is a universal phenomenon characteristic of all species of organisms from bacterial cells to higher plants and animals. The emission mechanisms are different: emission can take place in the process of respiratory exchange with the atmosphere or as a result of the activity of organs of internal secretion. It has been established that the emission of volatile organic compounds into the atmosphere increases markedly when various tissues of plants, particularly foliage, are damaged. Tissue damage occurs constantly, and some researchers are of the opinion that undamaged foliage virtually does not exist. Damage occurs when vegetation is eaten by phytophagous insects and herbivorous animals, when frost-cracks are formed, under the influence of wind, etc.

The functions of volatile biogenic organic compounds are varied and have not been studied in detail. Many of these compounds are the waste material resulting from the vital activity and toxic for the organisms producing them, whereas others are the means of intra- and interspecific interactions of organisms. The study of the life of insects provides examples of intraspecific interactions effected with the aid of chemical compounds emitted into the atmosphere. Volatile products of special glands, pheromones, serve for exchanging information: attractants ensure the search for individuals of the opposite sex, alarm pheromones cause flight or collective attack on the enemy, and aggregation pheromones lead to the accumulation of a large number of insects. Some compounds that control the number of organisms in a population keeping it in the ranges ensuring equilibrium with the environment have also been found.

Volatile organic substances are also used by many organisms for defence. The active emission of organic compounds by plant foliage in the first minutes after damage is considered to be a property developed in the course of evolution and ensuring nonspecific immunity from microbial infection. These compounds exhibit pronounced bactericidal and fungicidal properties. Hence, their emission prevents the penetration of microbes into tissues through damaged parts. The action of volatile emissions of plants can be directed not only against microorganisms but also against higher plants of other species. In this case they act most often as chemical inhibitors supressing the germination of seeds of competing plants. Hence, organic compounds emitted by plants into the atmosphere play a considerable role in the formation of vegetable associations. Furthermore, many volatile substances produced by plants serve for baiting or scaring away insects and other animals. These facts suggest that organic emission of plants are important to the formation of biocenosis as a whole.

At present, it is not yet possible to estimate even approximately the total emission into the atmosphere from the entire mass of living organisms. However, these attempts have already been made for individual biogenic sources. The above data on the distribution of the living substance of the planet between various reservoirs suggest a priori that the largest amounts of organic substances are emitted into the atmosphere from autotrophic vegetation, and the lowest amounts are emitted by animals.

3.1.1 Terrestrial Vegetation

Qualitative composition

The investigation of the composition of organic substances emitted into the atmosphere by plants is just beginning. It is hardly probable that this investigation will some day be carried out for all the representatives of the vegetable world because they are too numerous: flowering (angiospermous) plants alone number about 150 thousand species. It is noteworthy however, that although the species are extremely varied, the greatest specific weight in the total biomass is provided by a relatively small group of plants among those present in a given ecosystem. Thus, the forests of the boreal zone of Eurasia covering an extensive territory from the Atlantic to the Pacific Ocean mainly consist of several species of pine, spruce, larch and fir with a relatively small admixture of small-leaved trees, mainly birch and aspen, although the total quantity of species of trees in them is several hundred. The same can be said about the steppe exosystems. According to the data of Rice E. [233b] who studied the composition of herbaceous vegetation of the Oklahoma prairie, about 85% of the herbage consisted of only nine species and the rest was composed of other grasses of 20 different species. Cultivated landscapes exhibit even less varied species. The greatest sowing areas in the entire world are occupied by grain crops the most important of which are wheat, barley, corn, rye and rice. Hence, the general concept of the chemical composition of phytogenous organic components may be obtained on the basis of the investigation of emissions of a relatively small amount of edificating plants determining the main features of the structure of individual biocenoses. However, even this problem has not yet been finally solved: at present, only some data have been obtained on the composition of volatile emissions of the principal forest-forming species and a number of other plants growing under the canopy of coniferous forests of the Northern Hemisphere. Unfortunately, the information on the phytoorganic background of the atmosphere formed by the vegetation of highly productive forests of the tropical and subtropical zone is almost entirely absent.

Table 3.2 lists the composition of volatile emissions of some trees [69–71,

234, 235]. These trees belong to six families: Pinaceae (pine, spruce, larch and fir), Cupressacea (various species of juniper and thuja), Salicaceae (aspen, willow and poplar), Betulaceae (birch and alder), Rosaceae (mountain ash), Fagaceae (various species of oak) and Rutaceae (lemon and orange). Moreover, it also gives the composition of emission of some shrubs belonging to the family of Ericaceae (red bilberry and bilberry shrub and marsh tea) and mosses. These evergreen plants are an important component of vegetable associations of coniferous forests [71].

This list includes 78 compounds with the number of carbon atoms from one to ten. The qualitative composition of substances emitted by these plants is greatly varied: among them one can name paraffin, olefine and diene hydrocarbons, monoterpenes, saturated and unsaturated alcohols, aldehydes and ketones, esters, furan and its derivatives and chlorine-containing compounds. Most of these compounds with the exception of terpenes were considered to be anthropogenic pollutants because they were usually detected in high concentrations in urban air. However, these data show that they are simultaneously emitted by natural sources and, hence, are among the natural components of the atmosphere.

The composition of volatile substances is typical of each species but some features characteristic of individual groups of plants may be pointed out. A specific feature of coniferous trees is the emission of large amounts of $C_{10}H_{16}$ terpene hydrocarbons which usually make up more than 80% of the amount of the compounds emitted. In contrast, deciduous trees are characterized by a typically high content of high volatile components, although some of them (poplar, etc.) also emit terpenes. These compounds are also present in the foliage of a great number of herbaceous volatile-oil-bearing plants and are emitted into the atmosphere by transpiration.

One of the most often encountered substances in volatile emissions is

Table 3.2. Organic compounds in volatile emissions of arboreous plants and some plants growing under the canopy of coniferous forests

Compound	Plant species	Compound	Plant species
Methane	1	1-Octen-3-on	22, 23
Ethane	1	Dimethyl cyclopentenon	22, 23
Propane	1	Furan	6
Butane	1, 18, 19	2-Methyl furan	1–5, 12, 13, 20–24
Pentane	1–9, 18		
2-Methylbutane	1–8, 12, 18, 20–23	3-Methyl furan	1–5, 12, 13, 21
Ethylene	1	Ethyl furan	1, 2, 4, 8, 20–22, 24
Propylene	1, 8, 9		

Table 3.2. (*Continued*)

Compound	Plant species	Compound	Plant species
Butene	1–12, 15, 16	Vinyl furan	1, 2, 4, 8, 24
Pentene	1, 2, 5–9, 13–15, 20–23	Hexyl furan	24
Nonene	1, 6, 11, 16	Methyl chloride	14–16
Isoprene	1–17, 20–24	Chloroform	16, 18, 19, 23
2,3-Dimethylbutadiene	8, 13, 14, 24	Dimethyl sulfide	15, 22
p-Cymene	7–17	Santene	8, 11
Methanol	15	Cyclofenhene	8–10, 15
Ethanol	3–5, 10, 12–15 18–23	Bornilene	8–17
		Tricyclene	7–10, 23
3-Hexen-1-ol	1, 2, 5, 7, 8, 20, 21	α-Thujene	9, 10, 13, 17
Acetaldehyde	1, 3, 19	α-Pinene	3–19, 23, 24
Propanal	6, 22	β-Pinene	3–11, 15, 23, 24
Butanal	22	δ-Fenchene	8, 10, 12
Crotonal	17	ε-Fenchene	9, 12
Isobutanal	16	α-Fenchene	7–17
Benzaldehyde	19	β-Fenchene	7, 10, 12
α-Methylacroleine	1–4, 7, 9, 22, 23	Camphene	3–17, 23, 24
Acetone	1–24	Sabinene	12–18
2-Butanone	8, 13–15, 22	Myrcene	7–18, 23, 24
3-Butene-2-on	2, 4	3-Carene	7–17, 23, 24
3-Methyl-3-buten-2-on	22	α-Phellandrene	3, 8–11
2-Pentanone	6, 8, 15, 20–23	β-Phellandrene	8–11, 14
3-Pentanone	5, 7–10, 15, 16, 23	α-Terpinene	8–10, 17
		β-Terpinene	3, 8–10, 13–15
Ethyl acetate	9, 18, 19, 20, 21, 23	γ-Terpinene	9, 13–15, 17, 18
Ethyl propionate	18	Limonene	3, 7–18, 23, 24
3-Hexene-1-ol acetate	1, 4–8, 18, 20, 21	Terpinolene	9–17, 24
		Alloocymene	24
Methyl (α-methyl)butyrate	14, 18, 19	1,8-Cineol	3, 8, 14
Ethyl(α-methyl)butyrate	18, 19	Fenchone	16
Methyl(α-methyl)capronoate	14	Thujone	16
Diethyl ether	18, 23	Camphor	8
3-Octanone	22, 23	Menthane	14
		Anethole	24

*1- Willow, 2- Aspen, 3- Balsam poplar, 4- European oak, 5- European birch, 6- Sorb, 7- European larch, 8- European fir, 9- Scotch pine, 10- Siberian pine, 11- Silver fir, 12- Common juniper, 13- Seravshan juniper, 14- Pencil cedar, 15- Evergreen cypress, 16- Northern white cedar, 17- Chinese arbor vitae, 18- Lemon, 19- Orange, 20- Red bilberry shrub, 21- Bilberry shrub, 22- Fern, 23- Deciduous moss, 24- Marsh tea isoprene. Rasmussen R. A. has established that it is the most abundant among the components produced by the foliage of over 70 species of trees growing in the USA forests [72]. Isoprene evidently belongs to substances emitted by most plants, although in many cases its relative content is very low. Various species of willow, poplar, aspen and oak are among the trees emitting large amounts of isoprene, whereas birch, alder, mountain ash, maple and coniferous trees emit it only in small amounts [71, 234].

In addition to the compounds listed in Table 3.2, heavier compounds emitted by vegetation are also present in the forest air: $C_{15}H_{24}$ sesquiterpene hydrocarbons, $C_{20}-C_{33}$ n-alkanes, $C_{14}-C_{34}$ normal fatty acids, and $C_{21}-C_{33}$ alcohols. A specific feature of paraffin distribution is a relatively higher content of odd-numbered homologues: the value of the parameter CPI is much higher than unity. The maximum concentrations have been detected for n-$C_{27}H_{56}$ and n-$C_{29}H_{60}$ hydrocarbons. This distribution of homologues unequivocally indicates their biogenic origin. In plants, $C_{20}-C_{37}$ n-alkanes are present in the cuticular wax covering leaves and stems with a protective layer. The main components of wax are precisely odd-numbered homologues the highest concentrations of which are observed for C_{27}, C_{29} and C_{31} hydrocarbons. In contrast, in fatty acids and primary alcohols present in wax, even-numbered homologues predominate. It has been established that for n-paraffins in the cuticular wax of various grasses, the average value of CPI (if the concentrations of homologues from C_{12} to C_{35} are summed up) is approximately 9.2 [199]. This confirms the conclusions about the biogenic origin of higher paraffins, acids and alcohols in aerosols from remote rural areas and explains the reason for the increase in CPI in summer time during the active vegetative season.

The composition of volatile organic compounds emitted by cultured plants which occupy vast sowing areas has not been investigated in detail. Thus, until recently it has only been known that wheat emits into the atmosphere light unsaturated hydrocarbons: ethylene and butylene. The chromatographic-mass spectrometric analyses carried out by us showed that volatile emissions of cereals consist not only of light hydrocarbons but also of oxygen- and halogen-containing compounds. For example, the emissions of barley were found to contain pentane, hexane, acetone, ethanol, ethyl acetate, tetrahydrofuran, traces of methylene chloride and chloroform. Blooming buckwheat emitted C_5-C_7 hydrocarbons, diethyl ether, acetone, α-pinene and other terpenes. Volatile emissions of sugar-cane leaves (sample collection was carried out in Cuba by Perez R.) contained hydrocarbons (butane, pentane, hexane and pentene), acetaldehyde, ethanol, esters (methyl and ethyl acetate, ethyl propionate, ethyl-isobutyrate, methyl and ethyl isovalerates) dimethyl sulfide and traces of carbon disulfide.

Emission rate

The study of the emission rate of volatile organic compounds by foliage started as early as at the end of the 1920s by Nilov V. working at the Nikitsky Botanic Garden. He developed a procedure for the quantitative determination of terpenes emitted by coniferous plants, which included their absorption in sulfuric acid and oxidation to CO_2. The first results showed that the emission of organic compounds may be considerable: the crown of an individual specimen of arborescent juniper weighing 100 kg emitted during a hot summer day up to 30 g of terpenes [231].

A more systematic investigation by using various modifications of Nilov's method and mass spectrometry was continued in the 1950s by other Soviet authors. Bryantzeva Z., Artemjeva M. [232, 233a] and Sanadze G. [233c] determined the emission rates of organic substances by the foliage of more than 15 species of trees and shrubs directing great attention to the effect of various meteorological and biological factors: air temperature and humidity, illumination and vegetative phase, on the intensity of emission. Summing up the results of these investigations, it can be noted that the species of trees and shrubs investigated by these authors are characterized by the emission rates varying from 0.6 to $10 \mu g \, Ch^{-1}$ per gram of dry weight of foliage. The authors of refs. [232, 233a] detected a decrease in the emission rate of volatile substances at night time and after rainfall.

Specialists in the field of atmospheric chemistry directed particular attention to the volatile emission of plants after Went F. [236] had published the hypothesis that terpene hydrocarbons take part in photochemical transformations leading to the formation of aerosol particles. Model experiments carried out soon afterwards showed that terpenes and isoprene exhibit very high reactivity and under suitable conditions can very actively participate in atmospheric photochemical reactions with the formation of aerosols and ozone. Burkser E. et al. reported first in 1940 the positive correlation between the content of terpene and ozone in the air of pine forests of the Ukraine [79]. In this connection, the determination of emission rates of organic components by plants of forest-forming species covering extensive territories acquires particular importance.

For methodological purposes, it is possible to single out tentatively three groups of arboreous plants of the Northern Hemisphere: 1) coniferous trees emitting mainly terpene hydrocarbons, 2) deciduous trees emitting mainly isoprene, 3) deciduous trees the emissions of which contain little (or no) isoprene.

The first group is represented in the boreal zone of Eurasia and North America mainly by the plants of various species of spruce, pine, larch, silver fir and tsuga. Table 3.3 gives the data obtained for the last 30 years on terpene emission by these plants and by some species of plants of the tropical zone. It can be seen that the values of rates obtained by different authors with the application of different experimental procedures are in relatively good agreement. The rates averaged over a large number of measurements range from 1 to $35 \mu m \, g^{-1} h^{-1}$ at temperatures from 10 to 32 °C. One has the impression that in the case of plants of the tropical zone, the emission is higher, but it is premature to draw any conclusions on this subject.

However, it is possible to detect differences in rates for the representatives of the same family. Other conditions being equal, the highest values were detected

Table 3.3. Monoterpenes emission rates from various plant species

Plant species and location of observation	Air temperature, °C	Emission rates, $\mu g\,g^{-1}\,h^{-1}$ (dry weight)	Ref.
USSR			
Pinus Pallasiana	—	4.0–4.4	232
Pinus Pallasiana	22	6.4 ± 2.7[a]	233a
Pinus Montezumae	22	8.2 ± 1.6[a]	233a
Coniferous trees (5 species), including:	11–32	0.9–34.5	71
Picea excelsa	27	4.7 ± 0.5	
Pinus silvestris	20	4.2 ± 0.4	
Larix sibirica	27	5.4 ± 1.0	
U.S.A.			
Coniferous plants	—	8.9	237
Pinus taeda	20	2.2[b]	—
Pinus caribea	30	6.6	238
Coniferous trees (5 species)	15–30	1.1–26.5	239
Coniferous trees (5 species), including:	30	1.0–13.8	240
Pseudotsuga menziesii	19–32	5.4–15.6	
Picea excelsa	30	4.8	
Pinus silvestris	22–28	0.4–5.1	
Malaysia	30–35		241
Mallotus paniculatis[c]		1.02	
Hevea brasiliensis[c]		7.05	
Ficus fistulosa (fig)[c]		27.27	
Macaraunga triloba[c]		45.30	
Cuba[d]			
Pinus caribea	20	13.1–14.1	
Pinus caribea	23–26	23.4–37.4	
Pinus tropicalis	26	5.1	
Lemon	26–27	15.8–18.3	
Tangerine	26–27	21.4–23.5	
Orange	26–27	1.3–4.9	

[a] Fresh weight. [b] Zimmerman P.R. (1977). [c] Emission of isoprene and terpenes (sum). [d] Perez R.G. and Isidorov V.A. (1986)

by the present author for Scotch pine (*P. silvestris*) and Siberian pine. According to the data in ref. [240], the high rate of terpene emission is also characteristic of a North American plant, the Douglas fir. A much lower emission was detected by us for the European fir [71]. However, even for plants of the same species, considerable variations in the emission rates of volatile substances are observed. They depend on the meteorological conditions during sample collection and, mainly, on temperature.

There was contradictory information concerning the effect of the temperature of the environment on the emission rate of terpenes by coniferous plants. The authors of refs. [72, 240] found a direct relationship between these values, whereas the authors of ref. [239] did not detect this correlation. In order to elucidate the effect of temperature on the emission rate, the present author carried out experiments in the summer of 1981 with one specimen of pine according to a procedure described in ref. [71]. The rates averaged over 3 or 4 measurements are given below:

Temperature, °C	14	18	20	24	28
Emission rate, $\mu g\,g^{-1}\,h^{-1}$	1.6 ± 0.2	3.0 ± 0.3	4.1 ± 0.3	8.9 ± 0.9	18.8 ± 1.9

It can be seen from these data that the increase in the air temperature from 14 to 28 °C led to an increase of terpene emission by approximately one order of magnitude.

The absence of correlation in experiments reported in ref. [239] is probably due to the fact that experiments were carried out with different representatives of the pine family. However, even for plants of the same species considerable differences in emission rates of volatile substances can be observed as a result of different external conditions: the soil type, the level of subsoil waters, etc. Hence, it is very important to obtain averaged characteristics which smooth down these differences as far as possible. It was found that the temperature dependence of terpene emission determined by the author for plants growing in the subzones of the northern, middle and southern taiga (in Archangelsk, Vologda and Leningrad regions) and in the zone of mixed forests (in the south of Byelorussia) is

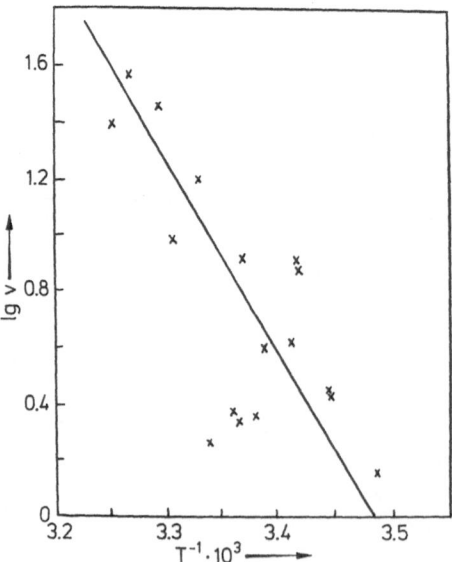

Fig. 3.1. Temperature dependence of the rate of terpene emission by pine needles

satisfactorily approximated by the equation described by a straight line in a system of semilogarithmic coordinates $\log V = a \times T^{-1} + B$. Fig. 3.1 shows the plot of the temperature dependence of the total emission rate of terpenes by pine needles from these areas. The same dependences were obtained for two other most important forest-forming species of the taiga forests of Eurasia: European fir and Siberian larch.

Protopopov V.V. et al. detected the existence of a distinct diurnal variation of volatile emission of plants [242a]. These authors determined oxidizability of air in vegetative chambers with young trees of various coniferous and deciduous Siberian species. The largest amounts of organic substances were emitted in the warmest time of the day. The same authors detected a change in the average values of air oxidizability during the vegetative season and concluded that the maximum emission of volatile organic substances takes place in June and the beginning of July, i.e., in the period of highest vegetative activity.

The second group including plants producing mainly isoprene is represented by a very great number of forest-forming species: oaks alone number about 450 species, and the trees and shrubs of the willow family number about 400 species. Isoprene is emitted in large amounts by the leaves of aspen, poplar, plane and acacia [71, 236, 239] and probably by many other plants. According to the data in ref. [239], the emission rate of seven plants (including five oak species) varied from 0.3 to 56.5 μg g^{-1}h^{-1}, depending on external conditions. The authors of ref. [241] recorded an even higher emission rate of isoprene (166 μg g^{-1}h^{-1}) for the foliage of *Elaeis guineesis* (oil palm) in Malaysia.

Rasmussen R.A. has established that isoprene is emitted by foliage only when illuminated: at night its biosynthesis and emission into the atmosphere stop [72]. However, it was reported in ref. [239] that the temperature dependence of emission exists but the correlation with the illumination level has not been observed.

In the summer of 1982, we carried out analyses of emissions of foliage of a 10

Table 3.4. Emission rate of isoprene from European oak (*Q. robur*) foliage depending on weather conditions [71]

Date of sampling	Time of day, h	Weather conditions		Emission rate	
		Temper-ature, °C	Cloudi-ness[a]	μg m^{-2}h^{-1}	μg g^{-1}h^{-1}
19.06	16^{30}	15	2	1616	12.4
24.06	15^{00}	19	1	3160	24.3
24.06	19^{00}	17	3	460	3.5
24.06	23^{30}	12	3	168	1.3
26.06	17^{00}	26.5	3	708	3.9
25.07	12^{00}	25.5	1	4700	36.2

[a] 1– sunny, 2– partially cloudy, 3– overcast

years old oak under various weather conditions. The results, given in Table 3.4, show that this dependence is complex: both factors (temperature and illumination) affect the emission rate of isoprene which usually represents about 80% of the sum of volatile components but the dependence on illumination is more pronounced. The same situation was observed for the foliage of aspen and bay-leaved willow, although the rate of isoprene emission by these trees differed greatly: under the same conditions the emission by the foliage of aspen and oak was higher by about one order of magnitude than that by willow leaves.

Flyckt D.L. et al. [239] determined the total emission rates of various organic components from ten species of deciduous trees belonging to the third group (emitting no isoprene). The general limits of rates were lower than for plants of the first two groups: they varied from 0.6 to 7.8 μg g^{-1}h^{-1}. These authors detected a linear temperature dependence of the logarithm of emission rate with the correlation coefficient of 0.88.

Emission scales

The first evaluations of the total emission of organic compounds into the atmosphere by arboreous plants were made by Nilov V. and subsequently by Artemieva M. The latter author on the basis of many hundreds of determinations of emission rates and the evaluation of the biomass established that 1 ha of a coniferous forest in the south of the Crimea emits on the average 4 kg of organic substances per day and 1 ha of a deciduous forest emits about 2 kg [233a]. Protopopov V. et al. later used the same approach [242a] and obtained similar emission characteristics for some forest types in the southern regions of Western Siberia. According to their calculations, the total terpene emission during a shortened vegetative period is 450–500 kg ha^{-1} for a Siberian cedar forest, 400–450 kg ha^{-1} for a pine forest (green biomass is 9–12 t ha^{-1}) and 200–220 kg ha^{-1} for a birch forest (the biomass of the foliage is 7–8 t ha^{-1}). My evaluation for the pine forests of the subzone of the middle taiga, carried out taking into account the experimentally found terpene emission rates and temperature variations during one day in the warm period, gave values from 1.6 to 3.2 kg ha^{-1}.

Bearing in mind that the Earth's surface covered with forests is 3620×10^6 ha [242b], it may be expected that the total emission of organic compounds into the atmosphere by the planet's forests is very substantial. The global evaluations started from Went F.'s paper [242c]. His calculations gave the value of 175 Tg y^{-1}.* Subsequently, Ramussen R.A. and Went F. [242d] on the basis of measurements of terpene concentrations in the air estimated their annual production to be 437 Tg. Zimmerman P.R. et al. (quoted in Ref. [237]) determined the emission rates of terpenes and isoprene by various plants and have extrapolated these data to the entire arboreous vegetation of the Earth. According to their calculations, about 396 Tg of isoprene and 545 Tg of terpenes (about 941 Tg in total) are emitted into the atmosphere annually [237].

---

*1 Tg $= 10^{12}$ g $= 10^6$ t.

Different approaches to the evaluation of global emission of organic compounds by vegetation are possible. One of them is based on the establishment of some parameters characterizing the vegetation as a whole. It would be interesting, for example, to use the ratio of the total productivity (increase in the biomass) to the fraction of organic substance emitted by plants into the atmosphere. Another parameter may evidently be the value of emission from unit area of the leaf surface averaged over many measurements. Analyzes showed that the typical values of terpene emission by coniferous trees of the boreal zone at temperatures of about 20 °C are close to 1 mg h^{-1} from 1 m^2 of surface area. For a number of broad-leaved trees of the middle zone, e.g., for oak and poplar, they are much higher and in day time attain $5–15 \text{ mg m}^{-2} \text{h}^{-1}$ [71, 238, 243]. Hence, assuming that the total foliage area is $644 \times 10^6 \text{ km}^2$ [244] and also considering that the emission rate of organic compounds is $1 \text{ mg m}^{-2}\text{h}^{-1}$ and is maintained on this level only for 12 hours a day at an average time of the vegetative season of 200 days, we obtain the value the global rate 1550 Tg y^{-1}. However, the initial value of the emission rate is taken to be that characteristic of plants of the boreal zone with relatively low productivity, whereas about 60% of annual production is provided by tropical forests (Table 3.1). Therefore the author is inclined to consider this emission rate, 1550 Tg y^{-1}, to be the lower limit of the real value of the emission of phytogenous organic compounds into the atmosphere.

However, this approach to the estimation of these values cannot be considered irreproachable: since the data are incomplete, the extrapolation of the characteristics obtained for individual types of vegetation and biocenoses to the entire vegetation leads to results the error of which cannot be determined. It is probably necessary to accumulate as precise data as possible for individual regions and then to sum them up.

This problem is also difficult: for the calculation of the emission of organic compounds by forests even in limited territories, many factors related to the structure of vegetation, the productivity of various biocenoses, and the climatic characteristics of the location have to be taken into account. We can take, as an example, the calculation of the emission rate of terpenes in the territory of one of the regions of the European part of the USSR: in the coniferous forests of the subzone of the northern and middle taiga covering an area of $0.626 \times 10^6 \text{ km}^2$.

These calculations required the determination of the temperature dependence of the emission rates of terpenes by the needles of two principal forest-forming species: spruce and pine. This dependence is expressed in the form of equations or plots similar to that shown in Fig. 3.1. It is also necessary to obtain data about the territories occupied by spruce and pine forests and about the percentage of admixture of deciduous species. Forests are divided into several types and the limits of needle biomass in them are different. The amount of needles also depends on the age structure of forests: they are at a minimum in young and overripe forests. For example, the investigations carried out in the European part of the USSR have shown that under the conditions of the northern taiga, bilberry pine forests occupy 50% of the area of all pine forests. Absolutely dry weight of needles in a 55- years-old forest of this type is 3.97 t ha^{-1} and the area of the needles is

Table 3.5. Area of coniferous forests, length of vegetative period and emission rate of terpenes in the zone of the Northern and middle taiga of European part of USSR

Region	Area covered with forests, 10^6 ha	Area of coniferous forests, 10^6 ha	Needle reserves, 10^6 t in pine forests	in spruce forests	Length of vegetation period, days[a] minimum	maximum	Emission rate, Tg·y^{-1} by pine forests minimum	maximum	by spruce forests minimum	maximum
Archangelsk region	20.1	18.3	37.8	169	120	155	0.15	0.59	0.09	0.16
Vologda region	9.2	5.7	19.8	42	153	163	0.12	0.34	0.04	0.06
Karelian ASSR	8.24	7.3	21.1	60	126	163	0.07	0.33	0.02	0.05
Kirov region	6.27	3.6	12.2	26	150	175	0.08	0.35	0.08	0.13
Komi ASSR	27.0	21.6	58.5	181	120	151	0.22	0.95	0.09	0.18
Perm region	8.59	6.2	10.3	61	130	168	0.07	0.29	0.03	0.06
Total	83.9	62.6	159.7	539	—	—	0.71	2.85	0.35	0.64

[a] Mean number of days per year with the mean diurnal temperature above 5 °C in the North (minimum value) and in the South (maximum value) of the region according to the data of 1880–1969 observations

$75 \times 10^3 \, \text{m}^2 \, \text{ha}^{-1}$. In the spruce forests of the same zone, absolute dry weight of needles attains $5.7–6.0 \, \text{t} \, \text{ha}^{-1}$ and their area is $130 \times 10^3 \, \text{m}^2 \, \text{ha}^{-1}$ [245].

The length of the vegetative period has also been taken into account with the application of data obtained for many years on the minimum and maximum number of days with the average diurnal temperature exceeding 5, 10, 15 and 20 °C. The evaluation of the above factors has made it possible to estimate that on this territory which makes up 11.2% of the USSR area covered with forest and only about 0.15% of the area of the Earth's forests, from 1.0 to 3.5 Tg of terpene hydrocarbous is emitted (Table 3.5).

It is interesting to compare the values obtained with the results of the last inventory of terpene emission by the USA forests [246]. The average biomass of needles in the USA forests is $5.6 \, \text{t} \, \text{ha}^{-1}$. The predominant amount of 6.6 Tg of terpenes is emitted by coniferous forests covering an area of $0.969 \times 10^6 \, \text{km}^2$. This implies that the emission of terpenes per unit area of the USA forests is 18% higher than the figure given by the upper limit for the European part of the USSR. If it is borne in mind that in our case forests with relatively low productivity growing to the north of 55 °N were taken into account, it should be recognized that the results of both evaluations do not contradict one another.

Hence, it seems that the evaluations of the emission rate of organic compounds by the Earth's vegetation exceeding $1500 \, \text{Tg} \, \text{y}^{-1}$ are not overestimated. This conclusion is of fundamental importance because it shows that on the global scale the kingdom of plants occupies the first place among the sources of atmospheric organic carbon. For making the evaluations still more precise, it is necessary to obtain data on the scales of emission by the vegetation of various bioclimatic regions of the Earth and in particular of the tropical zone.

In recent years, the ratio of the emission scale of biogenic components to their content in the atmosphere of rural areas has become the subject of animated discussion. Many researchers believed that the information about the high emission rate of organic compounds by plants on the one hand and their low concentration in the air of forests on the other are incompatible and contradictory. Among the possible reasons for this contradiction, Dimitriades B. [247] names the following factors:

1. The results of measurements of concentrations of natural volatile organic compounds are erroneously low.
2. The results of measurements of emission rates of natural volatile organics are erroneously high.
3. Investigations of organic components in the atmosphere of forests do not provide information about all the compounds emitted from plants.

As already mentioned, measurements of concentrations of organic compounds in the air of forests of the Northern Hemisphere and the emission rates by the foliage of arboreous plants characteristic of these forests carried out by different methods yield close values. The third of the above reasons is probably important. In the evaluation of the contribution of natural sources to the organic component of atmospheric air, only isoprene and terpenes are usually taken into account. However, the data listed in Table 3.2 on the composition of volatile emissions of various plants indicate that many other hydrocarbons and their derivatives which have traditionally been considered to be anthropogenic pollutants should also be taken into account.

Experiments show that meteorological conditions during sample collection also profoundly affect the concentration level of organic compounds under the forest canopy. The emission rate is directly dependent on the air temperature. However, the lowest terpene concentrations in the open atmosphere are usually observed in the middle of the day at the highest temperatures. The results of measurements carried out by the author at an interval of several hours in the coniferous forests of various regions of the European part of the USSR (Table 3.6) are characteristic in this respect. The highest terpene concentrations were observed in the morning and evening hours under the conditions of stratification of the atmosphere. The unusually high terpene concentration over a swampy area is due to the fact that here the main source of organic compounds is marsh tea, one of the most aromatic plants of damp northern forests and to complete calm during sample collection.

Very simple calculations carried out by a number of authors [80, 238, 243] in which the processes of turbulent diffusion are taken into account as well as model experiments carried out under natural conditions [248] show that the processes of dispersion of compounds emitted by plants profoundly affect their concentration level in the air. Hence, the contradiction between the emission rate and the concentrations of biogenic compounds reported above is only apparent. Under the conditions of the Northern Hemisphere, the concentrations

Table 3.6. Overall concentrations of monoterpene hydrocarbons in air under the canopy of pine forests [71]

Date of sampling	Location of sampling, wood type and age	Time of day, h	Air temperature, °C	Concentration, ppb
	1981			
03.08	Gomel region, pine plantation, 40 y	13–14	24	0.74
03.08	Same location	21–22	22	1.00
04.08	Same location	7–8	16	1.53
07.08	Kiev region, pine plantation, 20 y	8–9	21	4.40
07.08	Same location	15–16	28	1.58
	1982			
25.08	Vologda region, sphagnum bog with juniper and marsh tea	16–16^{30}	22	29.73[a]
26.08	Same location	16–16^{30}	20	3.49[b]

[a] Calm. [b] wind speed 1–$2\,\mathrm{m\,s^{-1}}$

of terpenes and isoprene exceeding 3–4 ppb are rather an exception than a rule, and their accumulation in large amounts is possible only when stable stratification of the atmosphere exists.

3.1.2 Aquatic Vegetation

As already mentioned, the emission of organic compounds into the environment is characteristic of all living organisms including those inhabiting water. As a result of phase transitions the mechanisms for which may differ, the components dissolved in water are emitted into air, and thus aqueous plant provide their contribution to the organic component of the atmosphere.

The qualitative composition of substances produced by aqueous plants is extremely varied. It has been established that sea and freshwater algae emit alcohols, aldehydes, carboxylic acids, alkylphenols, sulfur-containing compounds, terpenes and other hydrocarbons [194]. Particularly large amounts of organic components are emitted in the period of algae bloom of freshwater reservoirs, which causes a specific unpleasant odor. Gschwend P. et al. [249] investigated the composition of volatile organic components of seawater from the Peru upwelling region exhibiting high biological productivity. Some of the detected compounds were C_7–C_9 alkyl aromatic hydrocarbons, a series of normal aldehydes (hexanal, heptanal, octanal, nonanal and decanal), relatively large amounts of n-pentadecane and a compound with molecular weight of 108, to which the octatriene structure was tentatively ascribed. The highest concentrations of all compounds were observed in the surface layer of water. The average concentration of n-pentadecane in this layer exceeded $50 \, \text{ng} \, \text{kg}^{-1}$. Its source was phyto- or zooplankton. The anthropogenic origin of this hydrocarbon (e.g. by water pollution with oil products) is ruled out because the concentrations of the nearest homologues in the same waters did not exceed $2 \, \text{ng} \, \text{kg}^{-1}$. As to the origin of aldehydes, the authors have suggested that they could be formed as a result of autooxidation of some unsaturated carboxylic acids emitted by marine organisms.

The presence of relatively large amounts of n-pentadecane ($50–100 \, \text{ng} \, \text{kg}^{-1}$) and normal aldehydes in the surface layer of sea water has been confirmed by the investigations of Sauer T.C. [250]. He has identified 40 organic components. In the open ocean their total concentration was $80–210 \, \text{ng} \, \text{kg}^{-1}$, in the off shore waters of the Mexican Gulf it attained $320 \, \text{ng} \, \text{kg}^{-1}$ and in the water of an area subjected to a strong anthropogenic effect it was $1150 \, \text{ng} \, \text{kg}^{-1}$.

There are reasons to suppose that phytoplankton also serves as a source of light alkenes. For example, shortlived unsaturated hydrocarbons (ethylene and propylene) are constantly present in the air over the equatorial region of the Pacific Ocean [46, 62]. On the other hand, high concentrations of these hydrocarbons have been observed in the water from the upwelling region off the shore of South America, which may be associated with the high activity of biological processes in this region [62].

Marine vegetation emits into the atmosphere not only hydrocarbons, acids and aldehydes but also a number of other organic compounds among which, organic sulfides and halogen-containing compounds may be named.

Oceans are evidently the only large source of methyl iodide. The formation of CH_3I is considered to be related to the living activity of marine algae. Rasmussen R.A. et al. [251a] have carried out several hundred determinations of methyl iodide concentration both in the air over the seas of both hemispheres and in the water. The data obtained have been used as the basis of the evaluation of global emission equal to $1.3\,Tg\,y^{-1}$. It has been indicated that the largest amounts of CH_3I ($1\,Tg\,y^{-1}$) are emitted by sea shoals rich in marine vegetation.

Ocean water also contains and emits methyl bromide and chloride. According to the communication of Singh H.B. et al. (1979), the average concentrations of CH_3Cl in the surface layer of water in the tropical latitudes of the Southern hemisphere (9–25 °S) was $47.2 \pm 34\,ng\,l^{-1}$, whereas in the higher latitudes (30–39 °S) they were $6.4 \pm 5.6\,ng\,l^{-1}$. A more recent investigation by these authors [154] has shown that off the Californian shore (35 °51′N) the concentrations of CH_3Cl, CH_3Br and CH_3I in water were 11.5, 12 and $1.6\,ng\,l^{-1}$ and those in air were 633, 23 and 2 ppb, respectively. The depth profiles show that the upper ocean layer undergoing mixing is enriched with these halogenalkyls. The average flow of CH_3Cl into the atmosphere in the eastern subtropical part of the Pacific Ocean was $13 \times 10^{-7}\,g\,m^{-2}\,y^{-1}$ and that of CH_3Br and CH_3I was $1 \times 10^{-7}\,g\,m^{-2}\,y^{-1}$. The estimation of halogenalkyl emission by oceans obtained by extrapolating the above flow values gives the following values: $4.9\,Tg\,y^{-1}$ for CH_3Cl, $0.3\,Tg\,y^{-1}$ for CH_3Br and $0.3–0.5\,Tg\,y^{-1}$ for CH_3I [154].

The oceanic water is doubtless a large source of methyl chloride but its origin is not quite clear. Many workers believe that CH_3Cl is not a product of the vital activity of aqueous vegetation but is formed as a result of exchange processes between methyl iodide and chlorine ions:

$$CH_3I + Cl^- \rightleftharpoons CH_3Cl + I^-$$

In fact, methyl iodide is not thermodynamically stable in sea water, and kinetic calculations predict (and special experiments confirm) the possibility of these exchange processes.

The contribution of marine algae and phytoplankton to the atmospheric budget of sulfur compounds is very important. The interest in biogenic organic sulfur compounds is due to the fact that the troposphere and the stratosphere constantly contain sulfate aerosols profoundly affecting their radiation balance. The anthropogenic source of sulfur compounds, mainly in the form of SO_2, is well investigated. Its power has been determined with a sufficient precision on the basis of data on the world output and consumption of various kinds of fossil fuel. Careful inventory has shown that the anthropogenic source may be responsible for not more than 12.5% of the total sulfur content in the atmosphere. Even if such natural sources as volcanism and oceans emitting sulfate aerosols are taken into account, the deficit in the balance of sulfur compounds in the atmosphere still remains.

Therefore, many researchers have decided to consider biogenic sources for solving this problem. Indirect calculations make it possible to attribute to them over 50% of sulfur emitted in various compounds, which corresponds to the emission rate of sulfur of 60–90 Tg y^{-2} [130]. It was originally assumed that hydrogen sulfide formed in the bottom deposits of lakes, swamps and seas is the main biogenic component. However, hydrogen sulfide was found to be rapidly oxidized in the surface layer of sea water, and hence seas and oceans probably cannot emit it in considerable amounts.

Lovelock J.E. et al.[251b] suggested that the transition of sulfur from sea water into the atmosphere proceeds in the form of dimethyl sulfide, $(CH_3)_2S$ (DMS) exhibiting a combination of properties necessary for this transition: volatility, low solubility and higher stability with respect to oxygen dissolved in water than hydrogen sulfide. These authors have established that the waters of North Atlantic contain dimethyl sulfide. Rasmussen R.A. has established that many green, blue-green, red and brown algae of seas and freshwater basins emit dimethyl sulfide and dimethyl disulfide.

At present considerable experimental data have been gathered on the emission of sulfur compounds from oceans, and attempts have been made to generalize these data with the aim of describing their significance in the biogeochemical circulation of sulfur [253, 254]. These evaluations are very uncertain. For example, the total sulfur flow from natural sources is evaluated by the author of ref. [253] to range from 65 to 125 Tg y^{-1}.

Hundreds of measurements of DMS content in sea water have been carried out in the northern, central and southern Atlantic, in the tropical zone of the Pacific ocean and off the shores of Tasmania [129, 130, 254–256]. Analyses of more than 600 samples of sea water give an average DMS concentration of 102.4 ng l^{-1}, whereas in the equatorial regions of central Atlantic it attained

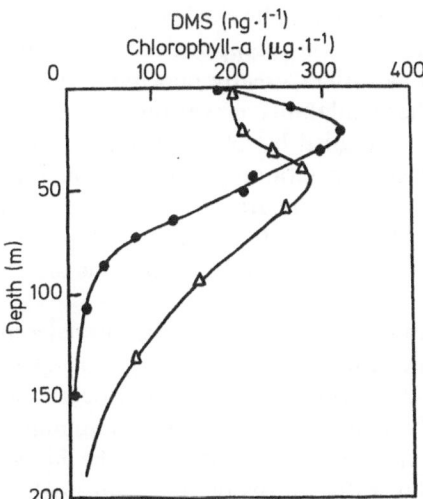

Fig. 3.2. Vertical distributions of DMS and chlorophyll-a in the central equatorial Pacific Ocean (0°00.2'S, 159°58.9'W) [255]

$740\,\mathrm{ng}\,\mathrm{l}^{-1}$ (based on sulfur). In the latitudinal range $18\,^{\circ}\mathrm{N}$–$6\,^{\circ}\mathrm{S}$, sulfur content was on the level of 80–$100\,\mathrm{ng}\,\mathrm{l}^{-1}$ and gradually increased to $430\,\mathrm{ng}\,\mathrm{l}^{-1}$ near the equator [255]. The vertical profile obtained in measurements carried out near the equator has shown that the maximum DMS content is observed at a depth of 20–40 m. The correlation between the concentration distributions of chlorophyll and DMS is an indirect evidence for the biogenic origin of the latter (Fig. 3.2). The authors of ref. [256] estimated the global rate of sulfur emission by the oceans in DMS to be 10–$18\,\mu\mathrm{g}\,\mathrm{m}^{-2}\,\mathrm{h}^{-1}$, and its total flow to the atmosphere to be 34–$56\,\mathrm{Tg}\,\mathrm{y}^{-1}$. More recent evaluations [130, 131] gave the emission value of $40 \pm 20\,\mathrm{Tg}\,\mathrm{y}^{-1}$.

3.1.3   Microorganisms. Qualitative Composition of Compounds Produced by Them

Microorganisms rank second (in mass) after plants as a reservoir of the living substance of the planet. Their role in the circulation of substances in nature as the destroyers of organic compounds is well known. Microorganisms mineralize any organic compounds of natural origin to carbon dioxide and water. These compounds serve for them both as sources of energy and sources of the construction material, carbon. However, complete mineralization of relatively complex organic compounds probably occurs as a result of combined activity of many destroyers. In other words, the molecules are gradually destroyed as they move along a kind of "conveyor" of the trophic chain consisting of many species of microorganisms. Each of them emits compounds which are the final products of their activity into the environment. A part of the volatile organic substances present among these metabolites leaves the trophic chain and passes into the atmosphere.

A characteristic feature of most microorganisms distinguishing them from higher plants and animals is their high adaptability. Their enzymatic systems can be rapidly reconstructed, and hence they can use quite varied substrates as sources of energy and carbon. Consequently, the composition of volatile compounds produced by microorganisms can differ greatly depending on the character of their environment and on the availability of various substrates. Therefore, it is hardly necessary to report a detailed list of compounds and to indicate all the species of bacteria, fungi, yeasts and actinomyces in the emissions of which these compounds have been found. These data are partly summed up in reviews, e.g. in ref. [257]. Hence, we will only enumerate the classes of organic substances and individual compounds detected mainly in the emissions of soil fungi and bacteria.

The final products of metabolism of many microorganisms are saturated and unsaturated hydrocarbons. C_1–C_4 alkanes are emitted by a large group of anaerobic soil bacteria. Many fungi produce olefinic C_2–C_4 hydrocarbons. Ethylene is detected among them expecially often and in predominant amounts. Among other unsaturated compounds, terpene hydrocarbons can be noted. Their emission is characteristic of many fungi destroying wood. There are no data in the literature on isoprene emission by bacteria and soil fungi although this emission is very characteristic of higher plants. However, the present author in collaboration with Zenkevich I.G. detected isoprene in small amounts among the volatile components of the fruit body of edible mushrooms.

Carbonyl compounds are represented by approximately 20 aldehydes and C_1–C_9 and ketones. In addition to the simplest unsaturated aldehydes, the emission of some unsaturated and aromatic aldehydes (2-hexenal, p-methoxy-benzaldehyde and furfurol) has been detected. Among ketones, some bifunctional compounds, diacetyl, acetoin and 4-methyl-2,3-pentadione, have also been found.

The C_1–C_8 alcohols are usually represented by isomers with a primary hydroxyl group. Among unsaturated alcohols, 1-octene-3-ol and cis-2-octene-1-ol are often encountered. Some fungi of the Fellinus genus produce benzyl alcohol.

Among carboxylic acids, normal C_1–C_7 homologues as well as isobutyric, isovaleric and acrylic acids have been detected. Esters are encountered in the emissions of many microorganisms in which almost all compounds formed by the combination of C_1–C_5 acids and C_1–C_5 alcohols of normal and iso-structure have been identified.

Sulfur-containing substances (methyl mercaptan, dimethyl sulfide, dimethyl disulfide and some thiophene derivatives) are emitted by very many species of bacteria of the Clostridium, Proteus, and Sarcina genera, actinomyces of the Streptomyces and Micromonospora genera, fungi of the Aspergillus, Fusarium, Candida, Schyzophyllum genera and other still unidentified species of soil micro-organisms.

A large number of C_1–C_7 primary amines and some secondary, tertiary, disubstituted (trimethylene diamine, putrescine, cadaverin) and fatty aromatic amines have been found among the metabolites of anaerobic bacteria – Clostridium, Proteus and E. coli.

It should be noted that in most cases volatile metabolites of microorganisms were studied after they had been isolated from soils and other natural environments by their cultivation on artificial substrates. The gas emission of soils themselves has not been extensively investigated. These data have been obtained only for a few compounds, e.g. for methane, its nearest C_2–C_3 homologues and sulfur-containing derivatives. It has not been elucidated, what are: the quantities of compounds of other classes emitted into the atmosphere from microbiological sources volatile components emitted by some soil micro-organisms do not always reach the atmosphere as a result of sorption and utilization by other microorganisms. There are reasons to assume that in a number of cases emissions attain considerable amounts. A communication has been cited above that in the air of the Alaska tundra unusually high con-centrations of n-butyl alcohol have been detected being on the average 190 ppb with a scatter of 36–445 ppb. A careful search for the sources of this compound has led to the conclusion that it is emitted as a result of the fermentation of vegetable matter by some species of bacteria of the Clostridium genus [12].

Almost all the above compounds are formed in the processes of degradation and biochemical transformation of organic substates: sugars, proteins and fats. In addition to this, microbiological processes of the transformation of non-volatile inorganic substances into volatile substances are known. For example, it has been established that fungi of the *Schyzophyllum* genus destroying wood can methylate inorganic sulfates with the emission of methyl mercaptan. Many soil microorganisms can methylate compounds of heavy metals: lead and mercury. The emission of dimethyl mercury has been detected in the collectors of sewage treatment plants. The natural emission of methyl chloride is probably also partially due to the ability of microorganisms to methylate inorganic chlorine compounds. The emission of this compound has been found for some soil fungi.

3.1.4   Microbiological Production of Hydrocarbons and Sulfur-Containing Compounds

Methane

Methane emission into the atmosphere occurs as a result of the activity of a group of anaerobic bacteria. They are the final link in the trophic chain of microbes decomposing complex organic compounds of cellulose type. Such simple substances as CO_2, hydrogen, methyl alcohol and acetic acid formed as a result of enzymatic degradation of biopolymers serve for methane-forming bacteria as substrates in methane synthesis which can be schematically represented by the following equations:

$$CO_2 + 4H_2 \rightarrow CH_4 + 2H_2O$$
$$CH_3OH + H_2 \rightarrow CH_4 + H_2O$$
$$CH_3COOH \rightarrow CH_4 + CO_2$$

Methane-forming bacteria inhabit swamps, the bottom of ponds and lakes, the silt deposits of the sea bottom, and soils. Some species of the methanobacterium genus are intensively reproduced in the digestive tract of higher animals (in the paunch of ruminants and in the blind gut of other animals) and some insects, where as a result of degradation of cellulose and other vegetable materials, carbon dioxide, hydrogen, organic acids and alcohols are formed. One of the most powerful sources of methane emitted into the atmosphere are swamps and temporarily flooded territories in which favorable conditions exist for the activity of anaerobic bacteria.

The evaluations of global methane emission have been carried out repeatedly, and some of them are given in Table 3.7. It can be seen from this table that considerable discrepancies exist in the evaluations of both total biogenic emission and its individual components. These discrepancies are caused by insufficient experimental data on CH_4 emission by various soils and animals under natural conditions. Consequently, in carrying out global evaluations the authors are obliged to use some kind of unfounded assumption.

The evaluations obtained by a direct transfer of the results of laboratory measurements of emission rates of CH_4 by soils and animals to natural objects

Table 3.7. Methane emission from natural and quasi-natural sources, $Tg\,y^{-1}$

Source	Evaluation according to references:			
	[15]	[258]	[259]	[260]
Swamps, lakes, the tundra, soils, seas	148–329	871	30–52	250–300
Rice fields	280	39	27–51	60–300
Animals	101–220	90	50–92	90–130
Termites	—	—	50–100	30–150
Total	529–829	1000	157–295	330–880

seem particularly open to criticism. The excessively rough division of all diverse biogeographical regions of the Earth into a limited number of ecosystems provides its contribution to the uncertainty of evaluations. For example, in ref. [258] all biogeographical regions are divided into 17 ecosystems many of which occupy areas exceeding $1 \times 10^6\,km^2$. For some of them, by the time of evaluation the data on the emission rate of methane were almost entirely absent. This refers, first, to tropical rain forests: the authors of ref. [258] had to assume for them as the "first approximation" the average rate of methane emission by soils of the damp subtropical forests of Florida. However, the investigations carried out later showed that the soils of tropical regions can serve not as sources but, rather, as a relatively considerable sink of atmospheric methane [261, 262].

In recent years, new information has been obtained on the emission of CH_4 by various biogenic sources: from coastal salt marshes, northern peatlands, the tundra, bogs, Alpine fen, etc. [263–267]. However, in our opinion this information is still fragmentary and insufficient for the correct evaluation of the global emission of this important component of the atmosphere. The solution of this problem will probably require the development of a system of planetary monitoring of the gas function of the soil cover [268].

In conclusion, we should consider the reasons leading to the increase in global methane concentration in the atmosphere. This effect is presumably directly or indirectly related to the influence of humanity on the environment. According to the authors of ref. [24], at present only about one third of the annual global methane production is indirectly controlled by the anthropogenic factor. The possible reasons for this may be due to the intensification of microbiological methane production as a result of favorable conditions for the activity of anaerobic bacteria due to the increase in irrigated agricultural areas, in areas occupied by rice plantations and in livestock. Statistical data reported by FAO UN show that all these factors are very important. For example the areas occupied by rice and livestock increased by about 20% in the period from 1970 to 1980.

Non-methane hydrocarbons

For a long time, the presence of methane homologues in soils was believed to be caused by their emission from deep accumulations of oil and gas. However, it has recently been found that in natural soils hydrocarbon gas is formed and consists of methane, ethylene, ethane and propane with a relatively high content of the two latter compounds [268]. The authors of ref. [269] believe that ethane and propane have a microbial source due to the biological formation of CH_4. Their gas chromatographic investigations of the composition of gases in soils of various groups in the middle of Russia and in the south of the Ukraine indicate that methane homologues are being constantly formed. Taking as an example a representative sampling of soils in the European part of the USSR, it has been established that the rate of ethane and propane ($F_{C_2H_6}$ and $F_{C_3H_8}$) emission by soils is linearly related to methane flow (F_{CH_4}). This relationship is satisfactorily described by the following regressive equations:

$$F_{C_2H_6} = -7.6 \cdot 10^{-4} + 7.8 \cdot 10^{-1} F_{CH_4}$$
$$F_{C_3H_8} = -5.5 \cdot 10^{-5} + 1.0 \cdot 10^{-1} F_{CH_4}$$

where F has the dimensionality $g\,m^{-2}\,h^{-1}$ [268].

These equations have been used by their authors for a preliminary evaluation of C_2H_6 and C_3H_8 flows from soils. In these evaluations, the duration of periods of biological activity (number of days per year with soil temperature above $10\,°C$ and with water supply in the soil above 1 to 2%) has been taken into account. The values obtained for the soils of the USSR and some other countries have been extrapolated to more extensive territories occupying about 80% of the area of soil cover on land. The integral emission of ethane and propane from these territories was found to be 46 to 101 and 9 to 23 $Tg\,y^{-1}$, respectively. This estimation must doubtless become more precise but its major significance is the demonstration of the role of soils as a large source of atmospheric ethane and propane.

Sulfur-containing compounds

The microbiological processes causing the formation of volatile sulfur-containing compounds in soils are varied. First, one should mention the aerobic decomposition of organic compounds containing cysteine, cystine and methyl cysteine. Many soil fungi and other microorganisms participate in this process. Sulfur compounds are also synthesized by a number of anaerobic bacteria using inorganic sulfates as a specific hydrogen acceptor in the oxidation of organic substances. Sulfides are formed in the course of this metabolic cycle. The reduction and methylation of sulfates has also been detected for some aerobic microorganisms.

In recent years, rather extensive experimental data have been obtained on the flows of sulfur-containing substances from microbiological sources on the surface of the continents [270–272]. However, they yield an extremely varied picture on the basis of which the main relationships of emission of these substances into the atmosphere are difficult to establish. First, both the composition and the emission rates of sulfur-containing substances are extremely variable. The spectrum of these substances was found to be much richer than had initially been assumed. Apart from hydrogen sulfide and dimethyl sulfide, it includes methyl mercaptan, dimethyl disulfide, carbon disulfide, carbonyl sulfide and a group of other unidentified components. From 1 to 10 peaks of sulfur compounds have been observed in the chromatograms of air samples collected in different locations [270]. The composition of these gases depends on many conditions: the content of organic substances and sulfates in soil, the acidity of soil water, the presence of metal ions capable of precipitating hydrogen sulfide and mercaptan in the form of sulfides, etc.

It was also initially assumed that the main amount of sulfur-containing compounds is emitted from exceedingly damp and salt marshes in tide-flooded coastal zones of oceans, whereas the contribution of continental sources, mainly freshwater swamps is slight. In fact, the highest emission rates are observed on the shores of warm seas and oceans [271] but detailed investigations carried out in recent years have shown that well drained dry soils of continental regions play the dominant role in the formation of the total flow.

Adams D.F. et al. [270] studied sulfur flows in various regions of the USA in the latitudinal range from 47 °N to 25 °N for four years. During this period, 760 samples were collected, and the average values of emission rates from soils were determined. These soils were tentatively divided into three groups: 1) tide-flooded damp coastal soils, 2) swampy soils with a high organic matter content located far from the sea shore and 3) dry continental soils. The first group occupied 7% of the region under investigation and emitted about 41% of the total sulfur content in volatile compounds. The second group occupied 18.7% of the area and its contribution was approximately 11%. The soils of the third group (74.3% of the area) were the most typical soils of the continents and produced about 48% of biogenic sulfur compounds.

These authors showed that emission rate increases from north to south and in the 47–25 °N range an exponential increase in flow from unit area is observed. They suppose that this phenomenon may be due to an increase in average annual temperatures, in the biomass of vegetation and the area occupied by swamps. On the whole, these data indicate the potential significance of damp tropical ecosystems, such as gilei (jungles, tropical rain forests) in the basins of the Amazon and the Congo and damp forests of south-eastern Asia in the formation of the global sulfur budget.

Extrapolating these data, Adams D.F. et al. [270] determined the global sulfur emission in the composition of volatile compounds from continental sources to be $64 \, Tg \, y^{-1}$. It is difficult to estimate the fraction of organic compounds in this flow because of very great variations in emission rates even from regions with a limited area. It has been reported [271, 272] that in the salt coastal marshes on the eastern shores of the USA, sulfur flow in dimethyl sulfide varies from less than 0.04 to $4.5 \, g \, m^{-2} \, y^{-1}$. Still greater variations (by a factor of 200) have been detected in this region by Adams D.F. et al., and the emission rates of methyl mercaptan varied from 0.0003 to $23.4 \, g \, m^{-2} \, y^{-1}$. The contribution of sulfur-containing organic compounds to sulfur emission in various locations of the USA ranged from 0.7 to 94%.

This varied emission picture makes it difficult to evaluate global sulfur emission in organic compounds. If it is assumed that the fraction of organic compounds in the sulfur flow is 16 to 17% for salt marshes, 5 to 6% for freshwater swamps and other exceedingly damp soils and 9 to 10% for dry soils, this emission is found to be about $6\,TgS\,y^{-1}$.

3.2   Geogenic and Other Natural Sources

In this section, data are reported on the emission of organic compounds into the atmosphere as a result of the degassing of the Earth's crust and mantle and during such large-scale natural phenomena as forest fires.

3.2.1   Volcanism

The contribution of volcanic sources to the formation of the organic reservoir of the Earth's atmosphere has not yet been taken into account in any attempts to understand the global cycle of carbon compounds. However, a number of investigations have shown that the processes of degassing of the upper Earth's mantle are accompanied not only by the emission of inorganic components, such as water vapor, carbon dioxide, hydrogen, etc., but also by that of a wide range of organic compounds. The first compound detected in the eruptive and fumarolic gases and in the gases emitted from the cooling lava rivers (streams) was methane. Its volume content in dried eruptive gases emitted during the eruptions of some Kamchatka volcanoes attains 7.4% [4a]. It has also been reported that in the fumarolic gases of some volcanoes of the Western Hemisphere its content varied from 0.08 to 0.4%. Apart from methane, volcanic gases always contain its C_2–C_6 homologues and light unsaturated hydrocarbons of the ethylene series.

The investigations of the pyroclastic material emitted during eruption: juvenile ash and volcanic bombs, have shown that it contains a complex mixture of low volatile organic compounds. For example, more than 150 components have been found in the solid products of eruptions of seven volcanoes on Kamchatka, the Kurile Islands and in Indonesia. The main fraction consisted of normal C_{15}–C_{36} alkanes, C_{18}–C_{36} isoalkanes and some polynuclear aromatic hydrocarbons see p. 66 among which pyrene, fluoranthene, benzo-(a)pyrene, benzo(ghi)perylene and coronene were identified. Among n-alkanes in the juvenile ash, C_{24}–C_{28} hydrocarbons were present at maximum concentrations. Moreover, there was no noticeable predominance of odd over even homologues. Apart from heavy hydrocarbons, ash and volcanic bombs contained their oxygen- and chlorine-containing derivatives [4a].

A complex mixture of organic compounds was detected in the particles of juvenile ash collected during the eruption of the Mount St. Helens volcano in the West of the USA in 1980. In addition to C_{15}–C_{32} alkanes and PAH, saturated mono- and dicarboxylic C_4–C_{10} acids, aromatic acids, aldehydes, ketones, alcohols and some chlorine-containing compounds (including polychlorinated biphenyls) have been identified [273].

Isidorov V.A. et al. have reported the results of GC-MS study on the composition of organic compounds in the solfataric gases of active volcanoes on the Kunashir Island [274–276]. The three active volcanoes on Kunashir (Mendeleev, Golovnin and Tyatya) are a part of the Kurile-Kamchatka volcanic arc. During the past 150 years, Kunashir was a scene of numerous volcanic eruptions, the last one taking place in 1973 (Tyatya). At present, all three volcanoes are in the stage of solfataric activity typical of most active volcanoes in the periods between eruptions.

Moreover, in 1987 the same authors investigated the composition of gases emitted by several Kamchatka volcanoes. On the Kamchatka peninsula there are 65 young volcanoes characterized by very high activity. Many of them are in the stage of continuous fumarolic or solfataric activity, and powerful hydrothermal systems function around them. Sample collection of volcanic gases was carried out in the crater of the Mutnovsky volcano and at some points of the Uzon-Geyser Volcano-tectonic depression.

In these samples collected on the Kunashir and Kamchatka in 1983–1987 the GC-MS technique identified 97 C_1–C_{12} organic compounds:

Propane	Octene
2-Methylpropane	Decene
n-Butane	Undecene
2-Methylbutane	Butadiene
n-Pentane	Pentadienes (2 isomers)
2,2-Dimethylbutane	Isoprene
2-Methylpentane	Hexadiene
3-Methylpentane	Methylcyclopentane
n-Hexane	Cyclohexane
2,2-Dimethylpentane	Methylcyclohexane
3,3-Dimethylpentane	Trimethylcyclopentanes (2 isomers)
2-Methylhexane	Benzene
3-Methylhexane	Toluene
3-Ethylpentane	Ethyl benzene
n-Heptane	*o, m, p*-Xylenes
2,2-Dimethylhexane	Styrene
2-Methylheptane	*n*-Propyl benzene
3-Methylheptane	Methyl ethyl benzene (2 isomers)
n-Octane	*sec*-Butyl benzene
2-Methyloctane	Cymene
n-Nonane	α-Pinene
n-Decane	Camphene
n-Undecane	Acetaldehyde
n-Dodecane	Propanal
Propylene	Butanal
Butenes (3 isomers)	Pentanal
Pentenes (2 isomers)	Hexanal

Hexene	Benzaldehyde
Acetone	Propylthiophene
2-Butanone	Methylmercaptane
Ethanol	Dimethyldisulfide
Isopropanol	Dimethyltrisulfide
1-Butanol	Acetonitrile
1,4-Dioxane	Acrylonitrile
Ethyl acetate	Crotonitrile (2 isomers)
Butyl acetate	Dichloromethane
Ethyl furan	Chloroform
Thiophene	Tetrachloromethane
2-Methylthiophene	Fluorodichloromethane
3-Methylthiophene	Fluorotrichloromethane
Dimethylthiophenes (3 isomers)	Difluorodichloromethane
Trimethylthiophene	Tetrachloroethylene
Ethylthiophene	Dimethyldifluoro silane

Of the compounds listed above, benzene was present in the highest concentrations. The analysis of air samples from Mendeleev and Tyatya volcanones collected in 1985 showed the benzene content to range from 2.7 to 3.5 mg m^{-3} (0.8 to 1.0 ppm). The total concentration of the C_4–C_{10} hydrocarbons in these samples was 3.5 to 4.1 mg m^{-3}.

It is of great interest that among highly volatile components halogen-containing compounds were detected, in particular, CFM's which had been considered to be solely anthropogenic atmospheric pollutants. Table 3.8 gives the concentrations of halocarbons detected in the air over solfataric fields and in volcanic gases. These data show that the concentrations of $CFCl_3$ and CF_2Cl_2 on the solfataric fields were usually 0.4 to 1.5 ppb which is 2.5 to 15 times those above the sea. In the case of chloroform and tetrachloromethane, this difference amounted to 1.5 to 2 orders of magnitude.

The volcanic gases sampled directly from solfataric vents contained much higher CFM concentrations reaching 160 ppb. High concentrations of these compounds were also found in gas bubble rising from the bottom of hydrothermal sources.

There are two viewpoints concerning the genesis of organic compounds emitted in volcanic gases into the atmosphere, the proponents of each of them usually disregarding the arguments of the other. According to the first of them, the "volcanogenic" substances are formed as a result of the decomposition of organic material contained in sedimentary rocks as lava and hot gases pass through them, as well as the destruction of vegetation and soil on volcanic slopes [273]. According to the second view, they originate from abiogenic synthesis from inorganic compounds [4a]. For instance, the formation of methane could be attributed to catalytic reduction of CO_2 and CO by hydrogen:

$$CO + 3H_2 \rightarrow CH_4 + H_2O; \quad CO_2 + 4H_2 \rightarrow CH_4 + 2H_2O$$

Table 3.8. Halocarbons content in the air of solfataric fields and in volcanic gases

Sampling site	Year	Subject of study	Concentration, ppb			
			$CFCl_3$	CF_2Cl_2	$CHCl_3$	CCl_4
Mendeleev volcano (Kunashir):						
Point 1	1983	Air	0.47	0.27	6.42	1.12
Point 2	1984	—"—	0.48–0.54	0.54	6.50	2.21
			0.62–0.87	0.42	—	—
Vent 1	1984	Volcanic gas	15–22.5	139.0	—	—
Vent 2			0.5–1.38	0.84	—	—
Vent 3	1985	—"—	60.5	11.2		
Vent 4			78.9	69.9		
Vent 5			59.4	140.3	41.4	3.8
Golovnin volcano (Kunashir):						
	1983	Air	0.63	0.39	6.16	2.34
	1984	—"—	0.79–1.48	0.39	—	—
Vent 1	1984	Volcanic gas	77.8–80.1	160.0	—	—
Vent 2			3.38–4.25	1.07	—	—
Tyatya volcano (Kunashir):						
side crater						
Point 1	1983	Air	1.02	0.74	—	—
Point 2			0.77	0.98	—	—
At the bottom of side crater	1985	—"—	34.9	30.2	31.0	—
At the bottom of central crater		—"—	23.3	16.7	16.6	0.71
Central crater rim		—"—	0.79–0.95	0.81–0.92	—	—
Mutnovskii volcano (Kamchatka):						
	1987	Volcanic gas	75.5	18.5	820.8	19.1
Caldera Uzon (Kamchatka):						
	1987	—"—	41.2	3.1	467.4	6.2
	1987	Gases from hydro-thermal sources	0.7–5.6	0.9–5.8	19.3–154	0.1–4.6

More complex compounds may be synthesized from methane and inorganic components in the ash-gaseous volcanic plumes with the participation of a fine-grained fluidized catalyst: juvenile ash. Bondarev V. B. and Porshnev N. V. have obtained evidence for the possibility of formation of a multicomponent mixture of organic substances by the interaction of methane and water vapor with red-hot volcanic lava [277a]. These authors found that a short (3 s) contact of a mixture of water vapor and methane with basalt heated to 1000 °C yields a mixture comprised of low molecular weight unsaturated hydrocarbons, acetaldehyde and acetic acid, benzene and a number of its homologues as well as polynuclear hydrocarbons ranging from naphtalene and diphenyl to pyrene. These temperatures are observed at the foot of the ash-gaseous volcanic plumes. It should be noted that under the actual conditions of eruption, very high temperature and pressure gradients are observed. The pressure differential of hundreds of atmospheres is the reason for the rapid removal of these compounds from the reaction zone thus ensuring their stability. Hence, the volcanic ash-gaseous plumes and clouds probably serve as gigantic chemical reactors in which various organic compounds are synthesized from simple gases.

Another argument for the abiogenic origin of these compounds is provided by the results of investigation of the isotopic content of carbon contained in them. Analysis of both extracts from ash and volcanic bombs and light C_1–C_3 alkanes and CO_2 shows that they contain a increased amount of heavy carbon. This feature markedly distinguishes volcanogenic hydrocarbons from biogenic substances present in sedimentary rocks. A correlation has also been observed between the content of hydrogen and helium of the mantle origin in gases of hydrothermal sources and the values of $\delta^{13}CH_4$: the content of ^3He and hydrogen of the mantle origin increases with the content of ^{13}C in methane from hydrothermal sources [277b].

These reasons cannot rule out partial inclusion of biogenic organics in the emission. Both processes, abiogenic synthesis and the decomposition of organic material, can apparently occur. The latter process is particularly essential in the early stages of eruption when the bulk of the organic material present in sedimentary rocks, soil and vegetation undergoes pyrolysis.

Analysis of solfataric gases emitted by the Mendeleev volcano on the Kunashir Island and the gases of hydrothermal sources of caldera of the Uzon volcano on Kamchatka showed the presence of terpenes (α-pinene and camphene) and isoprene in these gases. It may be assumed that these hydrocarbons are emitted when hot gases pass through the vegetable residues buried under the ash layer. However, volcanic gases contain components the biogenic origin of which is very unlikely. They include mainly halocarbons.

Chlorofluoromethanes are most probably secondary products of substitution of fluorine atoms for chlorine in CCl_4 molecules. Tetrachloromethane, in its turn, is produced in the substitution of chlorine for hydrogen atoms in the methane molecule. The CFM's are eventually formed in reactions similar to those used in industrial production of $CFCl_3$ and CF_2Cl_2:

$$2CCl_4 + 3HF \rightarrow CFCl_3 + CF_2Cl_2 + 3HCl$$

It is remarkable that the concentrations of these compounds in solfataric air and in volcanic gases (Table 3.8) are of the same order of magnitude, the variations in their content being similar to possible variations in the CFM mixture obtained from CCl_4 in industry depending on the actual reaction conditions.

In 1980, Rasmussen R. A. et al. reported the results of analyses of gases emitted during the eruptions of the Kilauea volcano on the Hawaiian Islands in September 1977 and November 1979 and those of fumarolic gases on the Mauna-Loa volcano [150]. In all samples, large amounts of methyl chloride were present: its concentrations exceeded the background concentrations by one or two orders of magnitude. However, these authors did not detect CFM's in the volcanic gases.

It is possible to find an explanation of the difference between the composition of volatile halogen-containing compounds if we consider the composition of gases in the volcanoes located in the transition zone from the continents to the oceans and that of gases in the oceanic and continental volcanoes. The difference between these three groups of active volcanoes is associated with different degassing stages of the Earth's mantle. Figure 3.3 shows a diagram of the composition of gases of the island volcano Kilauea (1), the Kluchewskaya volcano (2) forming a part of the Kurile-Kamchatka volcanic arc, and a continental Nyiragongo (3) located in Africa. Attention is drawn to the fact that the Kilauea and Nyiragongo gases do not contain hydrogen fluoride, whereas the gas of the Kluchewskaya volcano contains large amounts of HCl and HF. On the basis of these data one may expect the presence of CFM's in the emissions of volcanoes not only of the Kurile-Kamchatka arc, but also those of Japan, the Philippines, the island arcs and continental margins of America, etc. They represent 75% of the active volcanoes of the Earth.

The volcanoes of Central America attract particular attention as a source of CFM's. Recent investigation have shown that the emissions of Nicaragua

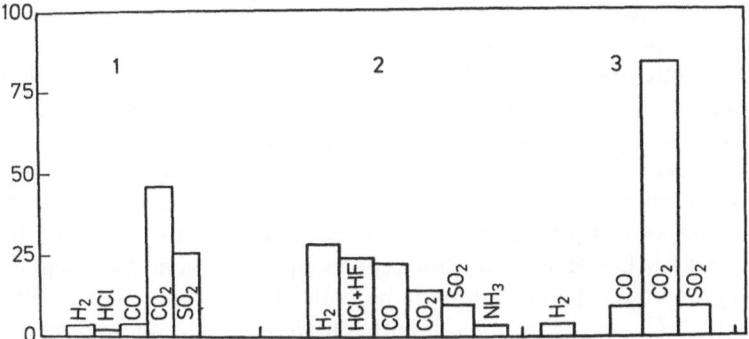

Fig. 3.3. Dependence of the composition of dried volcanic gases (vol %) on the type of the Earth's core, the composition and degassing stage of the mantle: *1*- oceanic mantle, *2*-mantle of transition zone from the ocean to the continent, *3*- continental mantle [4a]

volcanoes contain considerable amounts of HF (up to 0.14 mol.%) [278]. In fact, two air samples collected in a caldera of the crater of a Nicaragua volcano Masaya and transported to our laboratory contained CFM's in addition to other organic compounds. The concentration of $CFCl_3$ was 2.6 to 2.9 ppb.

The problem of whether CFM's are emitted during freatic eruptions or only in the stage of fumarolic and solfataric activity is still unresolved. Severe condition in the ash plumes and clouds may favor rapid decomposition of even these stable compounds. Unfortunately, up to the present analysis of eruptive gases was carried out only once during the eruption of Mount St. Helens (USA). There was no detailed analysis of volatile organic components but the authors of ref. [280] have reported that the gases contained considerable amounts of CH_3Cl and heavier halocarbons which remained unidentified. According to their data, $CFCl_3$ concentrations in the ash-gas cloud was slightly lower than the background concentrations.

It may also be considered to be established that CFM's are among the natural components of the atmosphere and their background content in the atmosphere before the beginning of the industrial era depended on the volcanic activity.

In this connection, comprehensive analysis of the air bubbles trapped in the cores taken from antarctic and arctic ices for the presence of CFM's would be of considerable interest. When studying methane concentration in such ices aged 1100 to 2600 y, Khalil M. A. K. and Rasmussen R. A. [279] detected $CFCl_3$ at a level of 17 ppt which is about 8% of its present-day content in the troposphere. The authors believed the $CFCl_3$ to have entered the samples in question with the ambient air. One cannot rule out the fact, however, that the chlorofluoromethane found by them is of a natural origin and, hence, reflects its content in the atmosphere thousands of years ago. If this assumption is correct, then investigation of arctic ice cores could yield additional valuable information on the volcanic activity and atmospheric composition in the past millennia.

It is difficult to evaluate accurately the levels of emission into the atmosphere of organic compounds from volcanic sources. Some data indicate that this emission is considerable and provides a noticeable contribution of organic carbon to the atmospheric reservoir, Ehhalt D. has evaluated the emission of methane from volcanic sources as $0.2\,Tg\,y^{-1}$ [15] but in the light of more recent data on the emission of hydrocarbon gases from hydrothermal fields and in eruptions, this figure may be considered to be only the lower limit.

Periodically active volcanoes are probably a considerable source of methane. According to Markhinin's calculations, not less than 0.47 Tg of methane and its nearest homologues were emitted into the atmosphere during the eruption of the Tolbachik volcano [4a]. The same author reports that the average content of organic carbon present in volcanic bombs of the Tolbachik volcano is 0.07%. The mass content of $C_{15}-C_{36}$ hydrocarbons and other high boiling compounds in the juvenile ash of various volcanoes attained 0.2%. If one proceeds from these values, then at a total mass of volcanic material erupted every year by more than 800 active volcanoes of the Earth of 3×10^{15} to 6×10^{15} g, the emission of organic substances should be about 2 to 5 Tg.

3.2.2 Emission of Organic Gases from the Earth's Crust

The Earth's crust contains gases in the free state, sorbed by various rocks and dissolved in water. The total amount of methane in sedimentary, granite and basalt layers is estimated to be 5.8×10^{19} g. A part of these gases reaches the Earth's surface along deep fractures and cracks and diffuses into the atmosphere. Particularly large amounts of hydrocarbon gases are contained in stratal waters of gas bearing basins and oil pools. For example, the saturation of the stratal water with gas in the megabasin of Western Siberia is 2 to $3 \, m^3 \, m^{-3}$ and in some cases even attains the value of $8 \, m^3 \, m^{-3}$ [281]. In some regions, the stratal waters enriched with methane reach the surface at such high flow rates that they can be used on an industrial scale.

Mud volcanoes most often encountered in oil-bearing areas are another source of gases rich in hydrocarbons. For example, the mud volcanoes near Baku erupt gases that contain 90 to 95% of methane with a small admixture of other light paraffins.

The existence of this "hydrocarbon breathing" of the Earth's crust is confirmed by the increased (sometimes as much as three times) methane content in the near ground air layer over oil-gas basins as compared to the background content [18]. It may be assumed that this high methane content should also be observed over regions rich in other types of fossil fuel: coal, brown coal and bituminous shale.

It has recently been established that in tectonically active regions, the increase in local methane concentrations as compared to the background concentrations in the overground air is an indication of a future earthquake. In a number of cases, CO_2 and CH_4 emitted from the fractures of the crust have been found to be enriched with ^{13}C isotope [282], which indicates that these gases are formed at a great depth. In our opinion, it may be suggested that the Earth's crust is a source of not only light hydrocarbons but also of many other organic compounds. In this connection, ref. [283] is of interest. Its authors estimated the anthropogenic emission of CF_4 to be about $28 \times 10^3 \, y^{-1}$, which is about 35% of the total emission of this compound in the atmosphere. The remainder of the emission evidently represents the fraction of an unknown natural source. These authors suggest that some geological formations may serve as a very weak source of CF_4 exhalations. In fact, fluorine organic and inorganic compounds have been detected long ago in gas inclusions of fluorine-containing minerals (fluorspar, etc.).

It may be suggested that gas inclusions of many minerals can also contain organic chlorine compounds. It is logical to expect the most active exhalations of these components mainly in seismically and tectonically active regions. In one of these seismic regions of the USSR (Kopetdag flexure near Ashkhabad) in the spring of 1988 we collected gas samples emitted by some numerous hydrothermal sources.* Gases from drill wells 2 to 2.6 km deep bored in the region of Ashkhabad deep fracture [282] were also collected. The results of preliminary analyses showed that spontaneous gases contained volatile hydrocarbons (CH_4, its homologues and alkenes) and halocarbons. Among the latter, $CHCl_3$, CCl_4 and $CFCl_3$ were found at concentrations of 2 to 3 mg m^{-3}.

The study of the Earth's crust as a source of atmospheric organic compounds is just beginning. Consequently, at present it is impossible to determine even approximately the power of this possibly important source of some organic gases.

3.2.3   Forest Fires

Forest fires often happen in summer time as a result of lightning or carelessness in handling fire: about 200 thousand cases of forest fires are recorded annually in the world. It is necessary to take this source into account for several reasons. First, fires extend over vast territories: according to the data of Robinson E., at the end of the 1960s forests burned annually on a territory of about 7×10^6 ha. Second, the combustion efficiency of damp vegetation is rather low: only about 20% of the vegetable material destroyed by fire is completely oxidized. The other part is emitted into the atmosphere in the form of volatile organic compounds or in solid particles (ash and condensation aerosol).

It has been established that the burning of wood and leaf litter is accompanied by the emission of large amounts of organic compounds, but their composition has not been investigated in detail. Many authors have identified over 50 volatile C_1–C_8 compounds in smoke gases. They contained methane and other C_2–C_7 paraffins, C_2–C_5 alkene hydrocarbons, benzene and its homologues, C_1–C_5 alcohols, C_1–C_7 carbonyl compounds including some unsaturated aldehydes and ketones. Furthermore, the condensates of smoke gases contain many phenol derivatives and polynuclear aromatic hydrocarbons [284–286].

Hence, combustion is accompanied by the emission of large amounts of various organic substances. Many of them rapidly take part in photochemical transformations, as indicated by the high ozone content in the smoke plume. These compounds are very reactive and therefore their lifetime in the atmosphere is short. However, in addition to them, during fires relatively stable compounds are emitted. These have a certain effect on the composition of the atmosphere not only on the regional but also on the global scale. This refers, first, to the emission of methane and methyl chloride.

Methane is one of the main gaseous products emitted during wood combustion under the conditions of oxygen deficiency. According to the evaluations carried out in refs. [283, 286], its total emission into the atmosphere as a result of processes of wood combustion varies from 12 to 110 Tg y^{-1}.

---

*The sampling and investigation of the composition of spontaneous gases was carried out by the author jointly with Yu.N. Fedorov.

However, it should be borne in mind that forest fires are only a short-term methane source, and at present there is no common opinion about the total effect of fires on its global production. During strong fires not only vegetation and leaf litter are burned out but also the organic material in soil. This leads to a long decrease in or even a temporary discontinuance of the activity of methane-forming soil bacteria. This deficiency of organic substances is probably partly compensated for by the fact that because of low combustion efficiency, a considerable part of vegetable residues returns into the soil with ash and undergoes anaerobic decomposition with methane evolution.

The emission of methyl chloride into the atmosphere is closely related to methane emission during the burning of wood or other vegetable material. The chlorine atoms contained in vegetable tissues serve as a trap for free radicals formed in the flames. It is assumed that during combustion methane is partly converted into methyl chloride. Chlorine content in wood pulp usually varies from 0.07 to 2.08 mg g^{-1} and is even higher in damp wood. Lovelock J. E. believes that about 2.2 mg of CH_3Cl is formed during the burning of 1 g of cellulose. According to the evaluation of the authors of ref. [283], the global emission of CH_3Cl formed during the combustion of vegetation ranges from 0.27 to 0.6 Tg y^{-1}. Hence, forest fires rank evidently second after the oceans as a source of the most widespread chlorine-containing organic compound in the atmosphere. It is assumed that both sources emit annually not less than 5.6 Tg of methyl chloride [287].

4 Man-Made Pollution

The activities of man have always influenced his environment. Even in very ancient times when man had very limited technological means, he caused important changes in the environment. For example, the use of fire for clearing forests with the aim of obtaining arable land and pastures as well as intensive cattle breeding have led to the extermination of vegetation and the appearance of deserts in vast territories. The accelerated development of technology greatly increased the influence of man upon nature. The period of scientific and technological revolution is characterized by a particularly rapid transformation of the environment involving changes not only in the vegetable and animal world but also in the chemical composition of the atmosphere.

Extensive use of fossil raw material for the requirements of the energetics and chemical industry is accompanied by the emission into the atmosphere of large masses of a variety of chemical compounds. The tremendous scale of this emission shows the possibility of many negative consequences involving threatening changes in the Earth's climate and the perturbations in the biological basis of human heredity. These consequences seem particularly menacing if the continuous trend towards urbanization is taken into account. According to the UN forecasts, by the year 2000 the population of the Earth will exceed 6.2×10^9, and about 3.5×10^9 will live in cities. Furthermore, population of some of these cities (New York, Mexico City, Tokyo, Rio de Janeiro, etc.) will exceed 20×10^6. It should be borne in mind that the most intensive industrial activities and the accompanying emission of atmospheric pollutants* are observed in cities.

Even now vast territories are affected by the city sources. Aircraft surveys of distribution of pollutant concentrations carried out in the 1970s showed that if the direction of motion of air masses to the leeward of large cities is stable, a single train is formed in which the distinctive features of individual sources are rapidly smoothed down. These trains lead to considerable pollution of air in rural areas even at distances of 150 to 200 km from the city. Moreover, the air of rural areas is often enriched with secondary products (aerosols, ozone, aldehydes, etc.) formed

---

*A pollutant is a substance present in the atmosphere in the amounts exceeding its natural background content and affecting man, the biosphere and materials.

Table 4.1. Emission rates of some anthropogenic impurities into the atmosphere

Pollutants	Principal sources	Emission rate, $Tg\,y^{-1}$
Hydrocarbons	Fuel combustion, chemical and petrochemical industry	97
NO and NO_2	Fuel combustion, fertilizer production	53
SO_2	Combustion of fossil fuel	130
CO	Combustion of fossil fuel	640[a]
CO_2	Combustion of fossil fuel	1.4×10^4
NH_3	Waste treatment	4
H_2S	Chemical industry, sewage treatment	3

[a] According to the data in ref. [237]

as a result of photochemical transformations occurring directly in the train of urban pollutants.

The scales of anthropogenic emission of organic compounds into the atmosphere have not yet been completely determined. Table 4.1 gives the evaluation of its greatest portion constituting the hydrocarbon fraction carried out in the middle of the 1970s. The table also contains data on the rates of anthropogenic emission of some inorganic components, which give an idea about the actual contribution of hydrocarbons to the total flow of impurities. In can be seen from this table that on the global scale the anthropogenic emission of organic substances is much smaller than the biogenic emission. However, in contrast to natural sources, anthropogenic emitters are concentrated mainly in cities occupying a limited territory. Hence, it is natural that relatively small anthropogenic emission of hydrocarbons and other organic compounds profoundly affects the quality of atmospheric air in cities and adjoining territories.

The modern trends of changes in the level of anthropogenic emission of atmospheric pollutants should also be borne in mind. At present, in all industrial countries laws have been passed on the protection of the air basin. For example, the emission of solid particles and SO_2 in the USA and Great Britain decreased by 50 to 70%. However, the control of the emission of hydrocarbons, nitrogen oxides and some other compounds has not led to tangible results: their content in the air of large cities is even increasing.

The qualitative and quantitative composition of atmospheric pollutants depends entirely on the character of individual sources and their contribution to the total emission. The most important sources of organic components of the atmosphere related to the industrial activities of man are automobile transport, industrial enterprises including electric power stations using fossil fuel, municipal services and agricultural enterprises. The contribution of each of these sources does not remain constant but varies with the changes in the type of industry in a given region.

4.1   Automobile Transport

4.1.1   General Characterization of the Source

In the first half of the 20th century, the principal amounts of the atmospheric pollutants of the cities of industrially developed countries were emitted with the exhaust gases of industrial enterprises. However, at present the automobile transport takes the first place among the sources of pollution.

According to various evaluations, in the USA its contribution to hydrocarbon emission is 53 to 63% [288]. It may be assumed that the contribution of automobile transport to air pollution will increase in spite of the measures taken for decreasing the toxicity of internal combustion engines. This results from the increasing number of motor cars: the rate of increase in the fleets of motor vehicles in the entire world exceeds that of the population. According to the forecasts carried out in Hungary and Czechoslovakia on the basis of the increase in the number of motor cars planned in these countries for the period from 1975 to 1990, the amount of hydrocarbons emitted with exhaust gases will increase by a factor of 1.7 to 2.4. However, even in the middle of the 1970s the contribution of automobile transport to atmospheric pollution was considerable. For example, in Hungary, 30 to 35% of the emission of pollutants consisted of exhaust gases.

In the Soviet Union automobile transport dominates all types of transportation. At present it transports about 82% of goods and more than 90% of passengers. The increase in automobile traffic leads to an increase in its contribution to atmospheric pollution. Even in the last decade in many cities of the USSR the amount of automobile transport emission was the same as that from other sources.

The emission of organic compounds by various assemblies of motor cars with spark ignition is distributed as follows: over 50% of the total amount of emitted compounds is provided by exhaust gases, evaporation from tanks and carburettors provides about 20% and leakage from the crankcase gives about 25% of hydrocarbons.

The crankcase gases are enriched with organic compounds as a result of partial leakage of fuel vapor from the engine cylinders. One of the reasons for the incomplete fuel combustion and the emission of hydrocarbons with exhaust gases is the extinction of the flame near the walls of the combustion chamber. The emission of hydrocarbons as a result of this process is considered to be about 9% of their content in exhaust gases. Another source of the unburnt hydrocarbons is the existence of small cracks in the walls of the engine cylinder partly sorbing the fuel.

The amount of organic compounds emitted with exhaust gases depends on many factors. The most important among them are the type of the engine and its technical state, the conditions of its operation and the fuel quality. The degree of hydrocarbon combustion is determined to a considerable extent by the fuel to air ratio in the fuel-air mixture. Gasoline engines with spark ignition operate at a fuel-air ratio close to stoichiometric, whereas compression-type engines (diesel engines) operate with excess air. The degree of compression of the fuel-air mixture is much higher in a diesel, although the maximum temperatures in both engines are similar and are about 2500 K. The combustion conditions in these engines differ, and hence the compositions of the components of exhaust gases are also different. Gasoline engines emit greater amounts of unburnt hydrocarbons and products of their incomplete oxidation (carbon oxides and aldehydes) than diesels.

The fuel-air ratio varies under various conditions of operation of the same engine. The most unfavorable combustion conditions when the air inleakage decreases and the mixture is excessively enriched with fuel, occur upon braking. The enriched mixture also enters the engine at no-load. Accordingly, the amount of unburnt hydrocarbons also changes. Investigations carried out in various countries show that their concentrations in exhaust gases of gasoline engines at no-load, at a cruising speed, at accelaration and braking are characterized by the ratio 2.3:1.0:1.3:9.2, respectively. This ratio shows that under the urban conditions the greatest amounts of pollutants are emitted with the exhaust gases when the cars stop for a short time at crossings [289].

Apart from the above sources of hydrocarbon emission into the atmosphere, there is another source also related to automobile transport. This is fuel losses in transportation and refuelling [290]. Before coming to the consumer, gasoline is poured from one container into another two or three times and is partially evaporated. According to some evaluations, during these operations, 1.4 Tg of hydrocarbons is annually emitted into the atmosphere.

4.1.2 Composition of Exhaust Gases

Exhaust gases are a multicomponent mixture consisting of not only the initial hydrocarbons but also the products of their incomplete oxidation, thermal decomposition and some other transformations. Variouis researchers have identified more than 500 organic compounds in the composition of exhaust gases. Hampton C. V. et al. [291a] published one of the most complete lists of volatile components comprising about 450 compounds.

The concentrations of individual compounds in exhaust gases vary over a very wide range. The simplest C_1–C_2 hydrocarbons are present in greatest amounts, but a great number of minor components, such as highly branched alkanes, alkyl naphthenes, polyalkyl substituted benzenes and naphthalenes, etc. have also been detected.

The average concentrations (in ppm) of over 75 C_1–C_{10} compounds belonging to the main components of the fraction of volatile hydrocarbons in exhaust gases of vehicles with a gasoline engine are given below (Dimitriades B. et al., 1968):

Alkanes		Isoprene	3
Methane	238	2-Methyl-2-butene	10
Ethane	16	3,3-Dimethyl-1-butene	1
2-Methyl propane	1	Cyclopentene	1
n-Butane	13	4-Methyl-2-pentene	1
		(*cis, trans*)	
2-Methylbutane	23	2-Methyl-1-pentene and	2
		2-Methyl-1-hexene	
n-Pentane	9	2-Ethyl-1-butene	2
Sum of hexanes	19	2-Hexene (*cis, trans*) and	1
Sum of heptanes	13	3-hexene (*cis, trans*)	
Sum of octanes	61	3-Methyl-2-pentene	2
Total	394	2-Methyl-2-pentene (*cis*),	4
		2,3-dimethyl-2-butene and	
		2,3,3-trimethyl-1-butene	
Acetylene hydrocarbons			
Acetylene	110	5-Methyl-1-hexene	1
Methyl acetylene	8	Cyclohexene	2
Total	118	5-Methyl-2-hexene (*cis, trans*)	—
		3,4-Dimethyl-2-pentene	3
		(*cis, trans*)	
Alkenes and cycloalkenes			
		1-Heptene	10
Ethylene	138	3-Heptene (*cis, trans*)	2
Propylene	54	Methylcyclohexene	3
1-Butene, 2-methyl	40	2,4,4-Trimethyl-2-pentene,	3
propene		3-methylcyclohexene and	
2-Butene (*cis, trans*)	8	4-methylcyclohexene	
3-Methyl-1-butene	1	2,4,4-Trimethyl-1-pentene,	1
		3-ethyl-2-pentene and	
1-Pentene	1	2-heptene	
2-Methyl-1-butene	5	1-Octene and 2-ethyl-1-hexene	1
2-Pentene (*trans*)	3	Other alkenes	3
2-Pentene (*cis*)	2	Total	310
Aromatic hydrocarbons		*o*-Ethyl toluene	4
Benzene	22	1,3,5-Trimethylbenzene	2
Toluene	107	1,2,4-Trimethylbenzene and	10
		tert-butyl benzene	
Ethyl benzene	6	*m*-Cymene and	2
m,p-Xylenes	13	1,2,3-trimethyl benzene	
o-Xylene	9	Other aromatic hydrocarbons	21
n-Propyl benzene	2	Total	206
m,p-Ethyl toluenes	8	Sum of hydrocarbons	1028

It can be seen from these data that methane, acetylene and ethylene make up more than 47% of the total amount of hydrocarbons. Unsaturated hydrocarbons constitute the most numerous group. If it is taken into account that acetylenes, alkenes and the C_1-C_3 alkanes are not contained in the liquid engine fuel, it is possible to estimate the contribution of the products of destruction processes to the formation of the hydrocarbon component of exhaust gases. It is apparently 67%. Unburnt hydrocarbons are represented by benzene and its homologues (20%) and the C_4-C_8 alkanes (about 13%). Hence, vehicle exhaust gases are found to be enriched with more reactive and toxic compounds than those present in the initial fuel.

These figures characterize the composition of gases emitted during the optimum conditions of engine operation. When it is run under the conditions of no-load and braking, greater amounts of unburnt hydrocarbons should be expected to be emitted.

A considerable amount of oxygen-containing compounds is formed as a result of partial oxidation of fuel hydrocarbons and products of their thermal decomposition. Seizinger D. E. and Dimitriadis B. [291b] investigated the composition of the principal components of exhaust gases of engines with spark ignition. The total content of volatile oxygen-containing compounds in automobile exhaust is much lower than that of hydrocarbons but these compounds exhibit high biological activity and are very deterimental to human health.

The concentrations (in ppm) of oxygen-containing compounds in automobile exhaust gases are given below:

Acetaldehyde	0.8–4.9	O-Hydroxybenzaldehyde	0.1–3.5
Propionaldehyde and acetone	2.3–14.0	Methanol	0.1–0.6
		Ethanol	0.1–0.6
Acrolein	0.2–5.3		
		2-Butene-1-ol and C_5H_7OH alcohol	0.1–1.1
3-Butene-2-on	0.1–42.6*		
Crotonaldehyde	0.2–7.0**	Benzyl alcohol	0.1–0.6
2-Butanone	0.1–1.0	Phenol and cresols	0.1–0.6
2-Methyl-3-propenal	0.1–1.0	2,2,4,4-Tetramethyl-tetrahydrofuran	0.1–6.4
2-Pentanone or 3-methyl-2-butanone	0.1–0.8		
		Benzofuran	0.1–2.8
3-Methyl-3-butene-2-on	0.1–0.8	Methylphenyl ether	0.1
		Methyl formiate	0.1–0.7
4-Methyl-3-pentene-2-on	0.1–1.5		
		Nitromethane	0.8–5.0
Benzaldehyde	0.1–13.5	C_4H_8O	0.1
Acetophenone	0.1–0.4	C_5H_6O	0.1–0.2
Toluyl aldehyde	0.1–2.6	$C_6H_{10}O$	0.1–0.2
Ethylbenzaldehyde	0.1–0.2		

*In a mixture with benzene
**In a mixture with toluene

In addition to the above derivatives, a number of nitrogen- and sulfur-containing compounds have been found in the exhaust gases of these engines. The presence of aceto-, propio-, and acrylonitrile and nitroethane was reported in ref. [292]. The authors of ref. [293] also found ethyl and dimethyl sulfide. The processes of evaporation and exhaust of unburnt fuel are also responsible for the emission into the atmosphere of organometallic compounds (tetraethyl or, in recent years, tetramethyl lead) used as antiknock additives to gasoline as well as for the presence of dibromo- and dichloroethane. The latter compounds are for preventing lead deposition on the internal surface of the combustion chamber. Ballschmiter K. and Mayer P. have found methyl chloroform, tetrachloromethane, tetra- and trichloroethylenes in the exhaust gases of internal combustion engines [294]. In contrast to gasoline spark engines, diesel engines emit a great number of heavy hydrocarbons and their derivatives. Among them, more than 100 C_{10}–C_{32} n-alkanes and their isomers, over 25 C_{10}–C_{22} alkylcyclohexanes and about 60 alkyl benzenes and C_{11}–C_{14} alkyl naphalenes have been detected. Most of these compounds are referred to a certain homologous series because it is impossible to carry out individual identification since the mixtures under analysis are very complex.

Polynuclear aromatic hydrocarbons and some of their oxygen- and nitrogen-containing derivatives present in exhaust gases of both types of engines attract particular attention. It has already been mentioned that the control of PAH content in the atmosphere is very important because many of them are carcinogenic. In recent years it has been shown that some PAH derivatives, in particular those the molecules of which contain carbonyl and nitro groups, are strongly mutagenic compounds.

PAH containing more than three condensed rings are not present in the fuel for combustion engines in any appreciable amounts. However, it is well known that the formation of condensed aromatic systems proceeds during the pyrolysis of light and intermediate oil fractions above 800 K. The same processes occur in the combustion chamber of engines. The reaction of cyclodehydrogenation can proceed according to the following scheme in which the formation of benzo(c)phenanthrene from ethyl benzene and naphthalene is taken as an example:

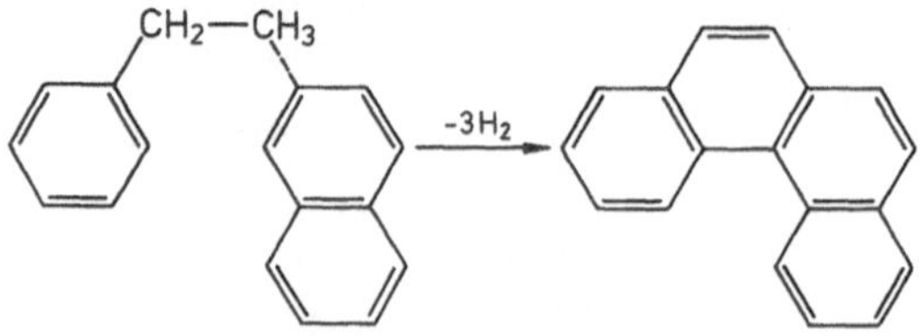

It has been established that PAH can also be formed as a result of more complex processes when the C_1–C_5 lower alkanes and cycloalkanes serve as the initial material. The total amount of compounds formed in exhaust gases is very great.

The partial oxidation of PAH leads to the formation of many derivatives. The scheme for the formation of aldehydes and ketones of the phenanthrene and fluorene series present in diesel exhaust gases in great amounts is given below [295]:

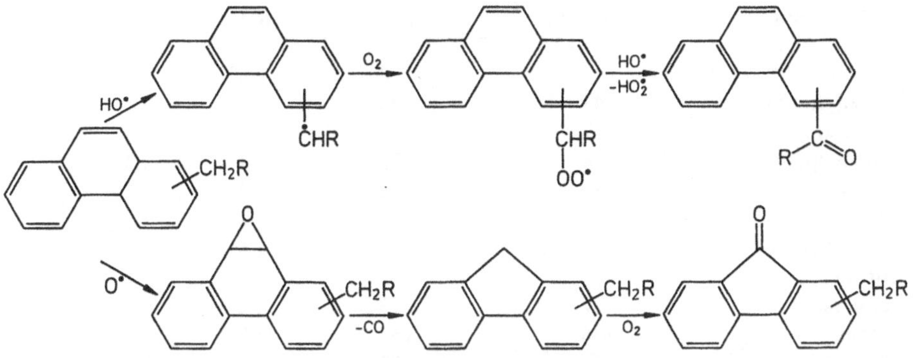

Automobile transport emits into the atmosphere in smaller amounts nitrogen- and sulfur-containing heterocyclic compounds with condensed rings: alkyl-substituted carbazole, dibenzo(cg)carbazole, benzo(def)dibenzothiophene and their derivatives, etc:

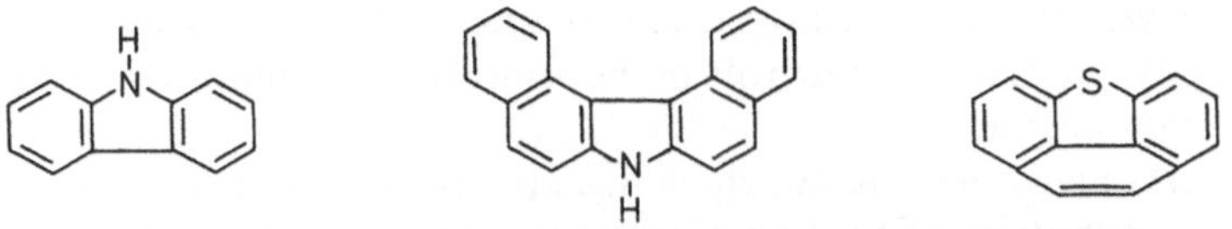

The initial material for their synthesis is provided by the compounds of the series of indole, quinoline, thiazole, benzothiophene and dibenzothiophene present in the wide motor naphtha and kerosene oil fraction boiling out in the 400 to 600 K range.

The nitration of PAH in engines evidently occurs by the mechanism of radical substitution with the participation of nitrogen dioxide. Isomeric nitroanthracenes, nitrophenanthrenes, nitrofluoranthenes and their homologues have been found in exhaust gases of diesel engines. The presence of many of these compounds and of 1-nitropyrene and dinitrosubstituted PAH (2,7-dinitrofluorene and 4,4'-dinitrobiphenyl) in soot particles emitted by diesels has been detected by Japanese researchers [296a]. According to their data, the average content of 1-nitropyrene in solid particles of exhaust gases is $8.6\,\mu g\,l^{-1}$ and that in the atmospheric aerosol of cities is about $21\,pg\,m^{-3}$.

Relatively little comparative evidence is available in the literature on the amounts of PAH emitted by diesels and engines with spark ignition. Shabad L.M. et al. (1981) (quoted in Ref. [297]) studied the exhaust gases of the motor cars most widely used in the countries of Eastern Europe and have concluded that the dependence of the qualitative composition and relative content of individual

Table 4.2. Average and relative concentrations of some PAH in Berlin's atmosphere

Hydrocarbon	Average conc., $ng\,m^{-3}$		Relative conc., %	
	Location with intensive traffic	Control points	in air	in exhaust gases
Benzo(a)anthracene	22	18	3.6	2.9
Benzo(ghi)perylene	47	20	5.6	5.1
Benzo(a)pyrene	17	11	2.5	2.8
Benzo(e)pyrene	28	25	3.7	3.3
Benzo(b)fluoranthene	29	25	4.9	4.6
Dibenzoanthracene	18	25	2.5	2.6
Indenopyrene	9	7	1.5	2.0
Coronene	14	12	2.4	2.8
Perylene	7	5	1.0	0.9
Pyrene	206	152	31.8	34.0
Fluoranthene	125	78	17.8	25.3
Chrysene	129	120	22.7	14.8

PAH on the engine type is very slight. However, Words K. [296b] reported that on the average the emission of PAH by gasoline engines is twice that of diesel engines. Nevertheless, all authors draw the unanimous conclusion that automobile transport plays the leading role in the atmospheric pollution of modern cities by these components.

A report of a research team which has investigated the PAH content in large cities of the countries of Eastern Europe is characteristic in this respect [297]. According to their data, the contribution of automobile transport to air pollution in Budapest, Leningrad and Berlin was 45, 54 and 60%, respectively. Table 4.2 lists the average and relative concentrations of some hydrocarbons detected in air in various locations of Berlin and in the exhaust gases of vehicles. The comparison of these values shows that the relative content of these components in air and in exhaust gases is similar. The authors indicate that the greatest relative contribution of automobile transport to the atmospheric pollution of Berlin by PAH is observed in summer time.

In order to evaluate the contribution of automobile transport to the pollution of the urban air, the characteristics of the relative composition of light hydrocarbons in fuel, exhaust gases and urban air have also been used. It has been proposed [96–98] to use the ratio of the sum of aromatic hydrocarbons to that of alkanes, r, as a parameter suitable to comparisons:

$$r = \left(\sum_{6}^{12} C_n H_{2n-6} \right) \bigg/ \left(\sum_{4}^{12} C_n H_{2n+2} \right).$$

Table 4.3 lists the results of calculation of group composition of C_4–C_{12} hydrocarbons in the air of seven large cities of the USSR. It also gives the

Table 4.3. Group composition (in %) of C_4-C_{12} hydrocarbons in the atmosphere of USSR cities, motor car fuel and exhaust gas

City	Collect-ion data	C_nH_{2n+2}		C_nH_{2n}		Aromatic hydrocarons	r
		Σ	n-alkanes	Σ	alkenes		
Baku:	19.5.77	35	16	37	14	28	0.8
	20.05.77	41	17	33	11	26	0.6
Tbilisi	24.05.77	48	22	18	7	34	0.7
Kemerovo	14.05.77	46	22	34	26	20	0.4
Tashkent:	19.07.77	46	23	20	8	34	0.7
	20.07.77	46	22	28	7	26	0.6
Irkutsk	11.07.78	51	25	18	7	31	0.6
Tashkent	03.11.78	54	25	12	2	34	0.6
Erevan	08.10.78	50	27	10	5	40	0.8
Leningrad:	04.05.77	41	24	21	11	38	0.9
	23.06.77	54	26	15	8	31	0.6
	04.07.77	56	24	10	3	34	0.6
	04.05.78	51	30	6	3	36	0.7
	18.05.78	58	28	7	2	35	0.6
	22.05.78	58	27	7	2	35	0.6
	25.05.78	55	27	7	2	38	0.7
	31.05.78	53	25	6	2	39	0.7
	06.06.78	55	26	7	1	38	0.7
	07.07.78	59	27	7	1	34	0.6
	0.4.08.78	57	26	6	1	37	0.7
	31.01.79	58	25	9	1	33	0.6
	29.03.79	56	24	12	2	31	0.6
	02.04.79	61	29	8	1	31	0.5
	24.04.79	61	30	10	2	29	0.5
Gasoline mixture		45	19	8	1	47	1.0
Exhaust gas of "Volga" car		24	11	13	12	63	2.6

composition of fuel obtained by mixing three gasoline grades in the proportions reflecting their average annual consumption in Leningrad at the end of the 1970s as well as the composition of the exhaust gas of the "Volga" car using the gasoline of the most typical grade.

The values of r in the atmosphere of these seven cities varied from 0.4 to 0.9, the average value being 0.6. This parameter is the maximum for exhaust gases because alkanes are burned more completely than aromatic hydrocarbons. On the whole, the group composition of hydrocarbons in the urban atmosphere of the USSR was closer to the composition of fuel than to that of exhaust gases. This may imply that one of the main sources of hydrocarbon emission into the air basin of the USSR cities is provided by loss of fuel due to evaporation. This indicates a certain difference from the character of air pollution in the cities of Western Europe and America for which a considerable resemblance between the compositions of atmospheric hydrocarbons and exhaust gases has been reported [44, 84, 100]. The parameter r calculated by the author according to the data in refs. [44, 84] was 1.1 for Houston and New York in the 1970s.

The American researchers use another approach for the determination of the contribution of individual sources to urban pollution by hydrocarbons. It is based on the fact that almost all acetylene is emitted into the urban atmosphere with the exhaust gases of automobile transport. The measurement of the ratio of concentrations of other hydrocarbons in air to that of acetylene and the comparison of this ratio with that characteristic of exhaust gases permits the evaluation of the contribution of the latter to total hydrocarbon emission.

According to the data in ref. [68], the ratio of the sum of C_5H_{10} alkenes to the concentration of acetylene in exhaust gases is close to 0.5, and the average ratio of the sum of concentrations of all hydrocarbons (with the exception of methane and acetylene) to that of acetylene is 15.5. The experimental values of this ratio for large cities of the USA (13.0 to 16.0) show that exhaust gases provide the main contribution to air pollution in these cities.

4.2   Industrial Production

The second important anthropogenic source of organic pollutants of the atmosphere of industrial countries after automobile transport are industrial enterprises. The amounts of pollutants emitted into the atmosphere are sometimes very great and attain several per cent of the mass of production.

The characteristic feature of this source is the variety of the emitted compounds. A particularly wide range of pollutants is emitted from chemical and petrochemical plants. Among these pollutants, many components of the initial raw material, intermediate, side and final products of synthesis are present. Thus, gaseous exhaust of plants producing fat substituents and synthetic detergents contain alkanes undergoing oxidation as well as intermediate and side products: carbonyl compounds, esters and carboxylic acids. Synthetic rubber plants pollute air mainly with initial monomers. The wood chemistry plants emit lower aldehydes, ketones, alcohols and C_1-C_6 carboxylic acids, many esters and terpenes. Pulp and paper mills emit not only hydrocarbons but also great amounts of gaseous malodorous substances, such as methyl-, dimethyl and dimethyldisulfides as well as formaldehyde, aliphatic and aromatic alcohols.

The total number of organic compounds emitted into the atmosphere by industrial enterprises probably comprises several tens of thousands. A detailed consideration of all these compounds and their sources is outside the scope of this book. We will deal only with the characterization of sources of some classes of organic compounds being of the greatest interest from the viewpoint of their influence on the chemical processes and the thermal conditions in the urban atmosphere as well as the compounds particularly harmful to human health.

4.2.1 Industrial Sources of Volatile Hydrocarbons

It was mentioned at the beginning of Chap. 4 that volatile hydrocarbons constitute the main fraction of organic compounds emitted into the atmosphere from anthropogenic sources. Industrial enterprises in industrial countries provide 20 to 25% of the total hydrocarbon emission. At the end of the 1970s in the USA approximately 26.7 Tg of hydrocarbons were emitted annually into the atmosphere. About 5.2 Tg of them were emitted by industry [298a]. The distribution of emission according to branches of industry in the USA is given in Table 4.4. The greatest contribution is provided by oil refining and the chemical industry (44.2%). Approximately the same percentage is provided by oil refining and petrochemical industries in the USSR.

Special investigations carried out at the plants of Union Carbide Corp. in the USA have shown that the main losses of volatile hydrocarbons at the plants of these industries occur at the pumping and compressor plants owing to leakage through seals of valves, flanges, etc. [298b]. These industries can be large local atmospheric pollutants.

Considerable amounts of volatile hydrocarbons are emitted into the atmosphere with exhaust gases of steam power plants using both liquid and solid fuel [299a]. The gases emitted during the combustion of liquid petroleum products contain aromatic hydrocarbons (benzene, toluene, ethyl benzene, etc.) and normal alkanes from C_7H_{16} to $C_{40}H_{82}$. According to the EPA data, when one ton of oil is burned at electric stations, 0.25 kg of hydrocarbons is evolved, and when one ton of coal is burned, 0.16 kg is emitted. The smallest amounts of hydrocarbons (0.48 kg per 10^6 m^3) are emitted when natural gas is used.

In 1975 the world production of natural gas exceeded 1000 Tg y^{-1}. According to some data, about 2% of methane, the principal component of this gas, is lost in extraction, transportation and use, i.e. more than 20 Tg of CH_4 is annually

Table 4.4. Emissions of hydrocarbons for point sources in the USA for 1977–1978

Industrial sources	Emission	
	thousand metric tonnes/year	%
Petroleum refining and related industries	1,240.0	24.0
Chemicals and allied products	1,039.9	20.2
Metal industries	415.5	8.0
Paper and allied products	275.1	5.4
Electric and gas service	246.2	4.8
Mining	176.3	3.4
Concrete and related products	79.1	1.5
Misc. sources	616.5	12.0
Other manufacturing	1,069.9	20.7
Total	5,158.5	100.0

emitted into the atmosphere [258]. According to the evaluation carried out by Ehhalt D.H. [15], anthropogenic sources are responsible for the annual emission into the atmosphere of 49 Tg of CH_4. About half of this quantity (up to 21 Tg y^{-1}) is lost by the gas works and 8 to 28 Tg y^{-1} is emitted during the extraction of coal.

It should be noted that this evaluation refers to the beginning of the 1970s and has probably become out of date. The scale of CH_4 emission in coal extraction is characterized by the fact that decontamination plants in the mines of the Donets basin alone emitted in 1984 about 5.7 Tg of methane.

4.2.2   Emission of Polynuclear Aromatic Hydrocarbons

Some branches of industry are large sources of emission of polynuclear aromatic hydrocarbons into the atmosphere. They are, first, by-product coke plants, petrochemical and metallurgical enterprises. The content of one of the most dangerous carcinogenic compounds, benzo(a)pyrene, at the emission sources on the territory of these enterprises in the USSR is given below:

Industry	Benzo(a)pyrene conc. in air, $\mu g\,m^{-3}$
By-product coke plant:	
coke shop	1.6. to 18.3
at the top of coke ovens	17.4 to 72.1
exhausts of quenching tower	7.7 to 18.7
coal tar processing plant	0.04 to 35.1
Coke gas plant	1.4 to 4.8
Coke pitch	up to 7150
Oil refining:	
paraffin plant	up to 0.03
bitumen plant	up to 0.2
coking plant	0.03 to 36.6
Asphalt concrete	up to 42.2
Ferrous metallurgy:	
blast-furnace	0.06 to 0.4
open-hearth furnace	0.02 to 0.04
cast-mold preparation	111 to 564.5
Aluminium production	42.9

These data show that particularly high local air pollution by PAH is caused by coal conversion to coke at temperatures of 1170–1270 K. The concentration of benzo(a)pyrene in coke gases is 187–673 mg m^{-3}. According to some calculations, its emission by various technological installations of a by-product coke plant with the output of $2 \times 10^6 \, t \, y^{-1}$ of coke including also the coal-tar processing plant and coked pitch industries is 2.3 to 7.5 kg·h^{-1}. The amounts of PAH in the atmosphere near the coke-oven batteries are particularly large during the charging and discharging of the oven chamber. At these moments they attain $72 \, \mu g \, m^{-3}$ [299b].

The emission of PAH is characteristic of industries using the products of coal pyrolysis: coke, resin and liquid oils. In ferrous metallurgy (in the blast-furnace and, to a lesser extent, in the open-hearth process), this emission is caused by both the release of PAH present in coke and their further formation in ovens at temperatures of up to 1000 K. The greatest amounts of these hydrocarbons are emitted into the air at the teeming of steel into molds previously lubricated with the coal-tar varnish consisting of 60% of anthracene oil and 40% of coal-tar pitch.

In the oil refining industry, the formation and high emissions of PAH are due to the processes of utilization of high-boiling products, mainly bitumens and stillage residues. In this case the main sources of carcinogenic substances are thermal cracking and coke production plants.

The high level of PAH emission in aluminium production is due to the use of anode paste also containing coal-tar pitch. Its sublimation during electrolysis leads to the isolation of large amounts of resin enriched with PAH. As a result, extensive territories adjoining aluminium plants are polluted with PAH.

Among other branches of industry greatly polluting the atmosphere with PAH, the production of soot (technical carbon) and thermal electric power industry should be named. Soot is used as a rubber filler, for example in the manufacture of tyres. The production of soot is mainly based on the pyrolysis of hydrocarbons. The average concentrations of PAH in the exhaust gases of these plants are approximately $160 \, \mu g \, m^{-3}$. Moreover, the quantity of carcinogenic compounds attains $13 \, \mu g \, m^{-3}$. These compounds are emitted into the atmosphere mainly as components of a longlived finely disperse condensation aerosol which is capable of penetrating deeply into the respiratory tract of man.

Electric power plants are evidently the most important source of PAH. At the beginning of the 1970s in the USA, they were responsible for the emission of about 500 t of benzo(a)pyrene annually, which represented about 38% of the total emission of this compound in the country. Power plants use various kinds of fuel: coal, residual fuel oil or natural gas. The investigation of flue gases of steam power plants has shown that the highest average concentrations of PAH of about $0.4 \, \mu g \, m^{-3}$ are characteristic of power plants using coal [300]. These authors have also established that the quality of liquid fuel profoundly affects the emission level of PAH: during a systematic investigation at one of the steam power plant, gas emissions with PAH concentrations varying from 0.02 to $2.55 \, \mu g \, m^{-3}$ have been recorded.

4.2.3   Halogenated Compounds

At present, many volatile halogen-containing hydrocarbon derivatives are synthetized in large amounts. At the beginning of the 1980s, the world production of only five compounds ($CHCl_3$, CCl_4, CH_3Cl, CH_2Cl_2 and CH_3CCl_3) exceeded $2.6 \times 10^6 \, t \, y^{-1}$ and is still increasing. For example, from 1960 to 1980 the production of methyl chloride increased from 63 to $360 \, kt \, y^{-1}$, that of methylene chloride increased from 93 to $570 \, kt \, y^{-1}$ and that of chloroform from 26 to $262 \, kt \, y^{-1}$. The production of chlorofluoromethanes is also increasing. The industrial production of CF_2Cl_2 and $CFCl_3$ began in 1935 and attained 473 and 399 kt, respectively in 1974. The total production of tri- and tetrachloroethylene was $2.06 \times 10^6 \, t$ in 1975. 1,2-dichloroethane is produced in greatest amount: $19.5 \times 10^6 \, t$ in 1975.

The main amount of methyl chloride produced in the world (70–80%) is used for the production of silicon materials and tetramethyl lead. About 10% is used for the manufacture of herbicides and methyl cellulose. Approximately half of the total amount of methylene chloride is employed as a solvent and the rest is used in the pharmaceutical industry and for the preparation of acetate fibers and films. The main consumers of chloroform and carbon tetrachloride are the processes of chlorofluoromethane production: about 60% of $CHCl_3$ is consumed for the manufacture of chlorodifluoromethane, and more than 95% of CCl_4 is used for the production of trichlorofluoro- and dichlorodifluoromethane. Methyl chloroform, tri- and tetrachloroethylenes are used almost entirely for the degreasing of metals, wool, cotton and articles manufactured from them. 1,2-Dichloroethane is the initial compound for the preparation of vinyl chloride (world production was $15.77 \times 10^6 \, t$ in 1979), methyl chloroform and tri- and tetrachloroethylenes. It is also used as solvent for dyes and varnishes. Chlorofluorohydrocarbons are widely employed as volatile components (propellants) in aerosol packing for dyes, insecticides, medicines, cosmetics, etc. About 85% of chlorofluorohydrocarbons are used for these purposes and only 15% are employed in refrigerators and air conditioning installations.

Among bromine-containing compounds, 1,2-dibromoethane was used in greatest amounts (about $180 \, kt \, y^{-1}$ based on bromine) in the 1970s. The production of methyl bromide was much lower (about $14 \, kt \, y^{-1}$ based on bromine); it is used as a soil fumigant for the extermination of insects and other agricultural pests in the soil. Bromofluorohydrocarbons CF_3Br, CF_2BrCF_2Br and CF_2BrCl manufactured industrially serve as fire inhibitors, but no reliable data on the scale of production of these fluorocompounds are available.

The specific conditions of using many volatile halogen-containing compounds are such that most of them are emitted into the atmosphere almost immediately. Table 4.5 lists data on the emission of halocarbons into the troposphere and their diffusion into the stratosphere.

Chlorofluoromethanes draw particular attention: less than 20% of the total quantity of chlorine in organic compounds is emitted into the troposphere in the form of $CFCl_3$ and CF_2Cl_2, whereas the chlorine fraction in the stratosphere exceeds 80%. This is due to the fact that trichlorofluoro- and dichlorodifluoromethanes are very inert and have virtually no sinks under the tropospheric conditions. Only in the stratosphere do they undergo photolytic dissociation with the isolation of atomic chlorine involved in the catalytic cycle of ozone

Table 4.5. Antropogenic emission of halocarbons into the troposphere and the stratosphere in 1978 [287]

Compound	Total emission into the troposphere, 10^3 t y^{-1}	Emission into the troposphere based on chlorine		Diffusion into the stratosphere based on chlorine	
		10^3 t y^{-1}	%	10^3 t y^{-1}	%
CH_3Cl	18	13	0.4	0.8	0.1
CH_2Cl_2	500	418	14.0	10.8	1.5
$CHCl_3$	12	11	0.4	0.3	0.1
CCl_4	50	46	1.5	46.0	6.6
CH_3CCl_3	500	399	13.4	39.9	5.7
$CHCl{=}CCl_2$	450	364	12.2	1.1	0.2
$CCl_2{=}CCl_2$	660	565	18.9	8.7	1.2
CH_2ClCH_2Cl	720	518	17.4	7.7	1.1
$CH_2{=}CHCl$	100	57	1.9	0.1	0.1
$CFCl_3$	400	310	10.4	310	44.5
CF_2Cl_2	450	264	8.9	264	37.9
$CHFCl_2$	50	21	0.7	8.4	1.2
Total	3892	2984	100	697	100

depletion. An opinion has been expressed that almost all the quantity of trichlorofluoro- and dichlorodifluoromethanes, 3.43 and 5.08×10^6 t, respectively (for 1975), should enter the atmosphere sooner or later.

The danger of important negative consequences of the unpremeditated anthropogenic influence on the protective ozone screen has led to a considerable anxiety among the scientists, the wide public and the government circles of many industrial countries. Many investigations have confirmed the existence of a potential danger of continuous emission of chlorofluorohydrocarbons into the atmosphere. As a result, in the middle of the 1970s a decision was taken in England about the restriction of the production and application of chlorofluorohydrocarbons and the introduction of more stringent requirements for controlling their emission into the atmosphere. In 1979 Sweden stopped the production and use of all kinds of aerosol preparations based on chlorofluorohydrocarbons. At the same time in the USA the use of $CFCl_3$ and $CFCl_2$ as propellants was prohibited. In the Federal German Republic the use of chlorofluorohydrocarbons was to be decreased by 50% in 1981 as compared to the 1976 level [301]. The result of this decrease in the level of production and emission was a certain decrease in the rate of growth of global background concentration of CCl_4, $CFCl_3$ and $CFCl_2$ [302].

At the same time scientists began the search for compounds capable of replacing trichlorofluoro- and dichlorodifluoromethanes as cooling agents and propellants without being so potentially dangerous to the environment. $CHClF_2$ and CF_3CH_2F are proposed as such substances. In contrast to $CFCl_3$ and

CF_2Cl_2, these halocarbons have a sink in the troposphere caused by their interaction with hydroxyl radicals.

$$CHClF_2 + OH^{\cdot} \longrightarrow H_2O + {}^{\cdot}CClF_2$$

It is assumed that as a result of this process only 40% of difluorochloromethane molecules can reach the stratosphere (see Table 4.5) and undergo photolysis with the abstraction of atomic chlorine. The molecules of 1,1,1,2-tetrafluoroethane do not contain chlorine atoms and do not participate in ozone depletion.

4.3 Municipal Sources

In the epoch of scientific and technological revolution characterized by rapid rates of urbanization of the population, the municipal services of large cities become a considerable source of organic atmospheric pollutants. Among the municipal plants emitting volatile organic compounds one can name dwelling houses and public buildings, plants for heat and water supply to the population and for dry cleaning, dumps, sewage purification plants, and the incinerators for solid waste. Although the contribution of the municipal services of cities to total anthropogenic emission is evidently not very high as compared to automobile transport and industry, there are reasons making it necessary to pay particular attention to this source. One of these reasons is that this is the source of emission of the principal amounts of some very dangerous longlived atmospheric pollutants. In this sense it is possible to speak about the role of municipal sources not only in local air pollution but also in the formation of the regional background of organic components of the atmosphere.

At present, this source is probably the least investigated. For example, there are almost no evaluations of the emission of pollutants by dwelling houses and public buildings. However, the investigations of ventilation emissions and emissions of refuse chutes carried out as early as in the 1970s permitted the identification of more than 40 toxic and malodorous substances. Mercaptans and sulfides, amines, alcohols, saturated and diene hydrocarbons, aldehydes and some heterocyclic compounds have been detected among them.

Natural gas used in housekeeping is a relatively pure source of energy but the products of its combustion contain 22 components [303]. In particular, considerable amounts of formaldehyde have been detected. Under normal conditions, up to 150 mg of CH_2O can be formed when 1 m^3 of natural gas is burned in the burner of the gas stove. Hence, formaldehyde content in ventilation emissions attains high values.

Other sources of malodorous substances in municipal service are the installations for the purification of non-industrial sewage water, in landfills of solid waste. Sewage water contains up to 0.025% of various organic substances, the main ones being fats, proteins and carbohydrates (totalling up to 0.02%) as well as fatty acids, amino acids and detergents. Moreover, large amounts of mineral components are present in it including salts of heavy metals. After settling and primary treatment (the removal of large particles of solid impurities), sewage water is introduced into installations for biological purification where bacterial degradation of organic components takes place. All purification stages usually taking about a week are accompanied by the evolution of malodorous volatile organic substances, mainly sulfur- and nitrogen-containing compounds. Some of the most harmful volatile components are organo-mercury compounds (CH_3HgCH_3 and CH_3HgCl) formed as a result of the microbiological methylation of inorganic salts of this metal. The processes of treatment and destruction of solid residues (treatment in autoclaves, drying, granulation, etc.) are also accompanied by emission of malodorous compounds.

In large city landfills, tremendous amounts (frequently over 10^6t) of garbage, waste, refuse and slag with high organic matter content are accumulated. Gas emission from landfills occurs as a result of both the evolution of volatile components when various materials gradually decompose and the processes of microbiological fermentation. It has been established that during waste storage, sulfur-containing compounds (methyl mercaptan, dimethyl sulfide, dimethyl disulfide and carbon disulfide), aromatic and unsaturated hydrocarbons, terpenes, alcohols and carbonyl compounds as well as methane (in the greatest amounts) are emitted [304–306a]. The total emission of sulfur-containing compounds from some investigated city dumps in the USA ranges from 0.01 to $0.26 \, gm^{-2}y^{-1}$, and the maximum quantity of volatile gases based on sulfur attains 0.2 kg per day. According to Vanni A. and Esposito A. [306b], one ton of waste in the landfills emits $25–30 \, m^3$ of methane for 25 years, and 70% of this quantity is emitted in the first ten years. Analysis of gas emitted from one of the largest landfills of Italy in the city of Brescia has shown that 52–56% of it consists of light hydrocarbons. Mercaptans concentration in this gas attains 21 ppm.

Large scale emission of malodorous substances from landfills and resulting air pollution near cities make it necessary to take measures for the neutralization of these gases. Systems for the collection and combustion of gases have been proposed. Moreover, since the territories for storage or burial of solid waste are limited, it is advisable to destroy them by burning in special incinerators. At present, in the whole world there are tens of thousands of such plants utilizing waste, and their number is increasing. It is remarkable that in these installations under the influence of fire many complex organic molecules undergo mineralization to simple substances, mainly water and carbon dioxide. It would seem that the circle is closed: for the first time man has unwillingly but deliberately taken upon himself the function formerly carried out almost exclusively by microorganisms; the return to various natural environments of the initial substances taken from them at different times by plants, animals and man himself. This is particularly necessary because industry manufactures a great amount of organic substances which virtually do not undergo biodegradation.

However, the procedure of burning solid waste is not yet sufficiently perfect and does not ensure complete destruction of all organic matter. Moreover, the gases emitted from the incinerators destroying waste contain dangerous pollutants formed or liberated upon combustion. More than 80 compounds of various classes detected in the emissions of municipal plants destroying solid city waste are listed below [307].

$C_{13}H_{28}$–$C_{30}H_{62}$ n-Alkanes	Biphenyl
Naphthalene	Acenaphthylene
1-Methylnaphthalene	Anthracene
2-Methylnaphthalene	Phenanthrene
Dibenzoheptafulvene	Tetrachlorobenzene (2 isomers)
Fluoranthene	Pentachlorobenzene
Pyrene	Hexachlorobenzene
Benzofluoranthene	Chloronapthalene
Benzopyrenes (2 isomers)	Pentachloronapthalene (3 isomers)
Perylene	Hexachloronaphthalene
Triphenylene	Tetrachlorobenzofuran
Dibenzofuran	Chlorobiphenyl
Fluorenone	Tetrachlorobiphenyl
Anthraquinone	Pentachlorobiphenyl
Xanthone	Tetrachlorodibenzodioxine (3 isomers)
Dicyclohexyl adipinate	Pentachlorodibenzodioxine (2 isomers)
Dibutyl phthalate	Hexachlorodibenzodioxine (6 isomers)
Benzylbutyl phthalate	Heptachlorodibenzodioxine (2 isomers)
Dioctyl phthalate	Octachlorodibenzodioxine
Trichlorophenol	Tetrachlorodibenzofuran
Tetrachlorophenol	Pentachlorodibenzofuran (5 isomers)
Pentachlorophenol	Hexachlorodibenzofuran
p-Chlorobenzoyl chloride	Heptachlorodibenzofuran
Chloracetophenone	Octachlorodibenzofuran

It is noteworthy that among these compounds polychlorinated biphenyls, dibenzofurans and dibenzo-p-dioxines are present:

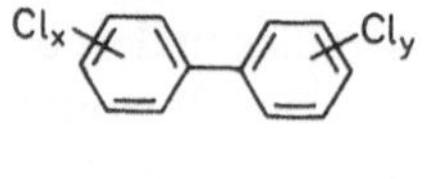

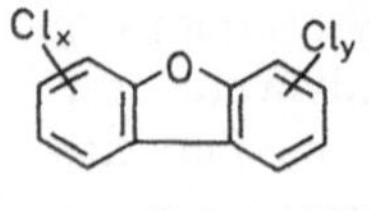

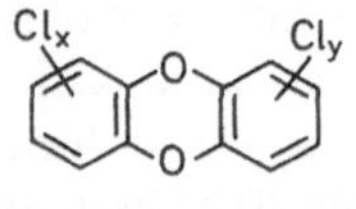

The compounds of these three classes are strong toxins very stable in the environment and capable of bioconcentration: accumulation in the fatty tissues of the organism.

Polychlorinated biphenyls (PCB's) have been produced by chemical industry since 1930, and up to the present their output was approximately 1×10^6t. Their application is based on chemical inertness, incombustibility, thermal stability, low volatility and high dielectric constant. About 100 PCB's are used in industry.

Most PCB's (56–65% of the world production) are used in the electrotechnical industry as a dielectric in capacitors and transformers. Up to 30% are employed as plasticizers for carbon paper, plastics and dyes, and the rest are used in heat exchangers and hydraulic drives.

The pollution of the environment with PCB's is considered to take place not during their production but during their further use and removal as waste, in particular in combustion. Highly chlorinated PCB's with a chlorine content of about 60% (e.g. arochlor 1260*) are particularly stable to pyrolysis. Entering the atmosphere PCB's are adsorbed on suspened particles the lifetime of which is two or three days. In various regions of the Earth, PCB concentrations range from 0.05 to 50 ng m^{-3}. Their content in the air over the north-western shore of the USA is about 5 ng m^{-3}. It should be noted that just as chlorofluoro-hydrocarbons, PCB's are not purely anthropogenic pollutants of the atmosphere. Isomeric pentachlorobiphenyls have been detected by Pereira W.E. et al. [273] in the products of a volcanic eruption.

Since PCB's are very toxic, many countries stopped their production at the beginning of the 1970s. Nevertheless a great amount of them is still in use. For example, according to some evaluations, about 35 kt of PCB's were used in the USA in the first half of the 1980s mainly in electric equipment.

Polychlorinated dibenzo-p-dioxines (PCDD's) and dibenzofurans (PCDBF's) are also emitted into the atmosphere with flue gases and as a part of volatile ash from garbage-disposal incinerators [308]. 75 isomeric PCDD's and 135 PCDBF's are known. The most toxic PCDD's exhibiting teratogenic, mutagenic and carcinogenic properties are 2,3,7,8-tetrachloro-, 1,2,3,7,8-pentachloro-, 1,2,3,6,7,8-hexachloro- and 1,2,3,7,8,9-hexachlorodibenzo-p-dioxines. According to the data of various authors, the total PCDD content in volatile ash is 180 to 2060 ng g^{-1} and that in flue gases attains 1540 ng m^{-3}. More or less the same concentrations are also characteristic of not less toxic PCDBF's (1300 ng g^{-1} and 1320 ng m^{-3}) [309].

It is universally recognized that polychlorinated dibenzo-p-dioxines and dibenzofurans are formed in combustion processes. Their precursors are phenol compounds and polyphenols always present in vegetable residues (e.g. in the form of lignin).

It is not surprising, therefore, that forest fires were found to be one of the most important sources of these compounds. For example, according to ref. [310], forest fires are one of the largest sources of PCDD's and PCDBF's on the territory of Canada. Here we observe again the fact of the natural origin of "exclusively anthropogenic" atmospheric pollutants.

Recently it has been reported the PCDD's and PCDBF's are emitted into the atmosphere with exhaust gases of motor cars using ethylated gasoline with

---

*The first two figures in the four-digit code refer to the hydro-carbon backbone (biphenyl) and the last two denote the mass content of chlorine in the mixture. Arochlor 1260 is a PCB mixture containing 60% chlorine.

dichloroethane addition. The authors of ref. [311] have concluded that under the conditions prevailing in Sweden, motor cars can emit the same quantities of these compounds as are emitted from 2–20 incinerators of average power.

According to the data in ref. [312], the content of polynuclear aromatic hydrocarbons in volatile ash fraction from garbage-disposal plants attains 1.1 μg g^{-1}. The formation and emission of PAH in the condensed phase and in the form of vapor is characteristic of all installations for burning fuel. Particularly large amounts of these compounds are emitted by small boiler rooms: the combustion of all kinds of fuel in them is accompanied by the emission of greater amounts of PAH than in large power-plants at the same standard of fuel consumption. The increase in PAH concentrations in winter time reported everywhere [194] is doubtless due to the beginning of the heating season. In the Northern Hemisphere, the maximum PAH content in the atmosphere is observed in the period from December to March. Moreover, the greatest amplitude of seasonal variations is usually observed in relatively small towns in which it is not so greatly smoothed down by the constant contributions of automobile transport and industry which do not depend on the season.

5 Chemical Transformation of Organic Compounds in the Atmosphere

One of the main task of atmospheric chemistry is the description of global balances of chemical compounds present in the atmosphere. The peculiar features of the balance of a given component affect its spatial distribution, determine the significance of its contribution to the radiation conditions and hence the dynamics of the atmosphere, influence the hygienic characteristics of air, etc.

In order to determine the completely balanced budget of a component, it is necessary to know all its sources and the mechanisms of its removal called sinks. The principal characteristic of a component budget is the time of its residence in the atmosphere (τ). The residence time implies the period of time from the emission of the component into the atmosphere to the sink from it. This value is different for different molecules, and hence an averaged characteristic including the probability distribution function $p(\tau)$ of the residence time of molecules of a given compound is used. The average time of its residence in the atmosphere is determined by the equation.

$$\bar{\tau} = \int_0^\infty \tau \, p(\tau) d\tau \tag{5.1}$$

It should be noted that the values of τ for the same component in the troposphere and the stratosphere and often in different hemispheres are different. Hence, this value is meaningful only when applied to a definite atmospheric reservoir.

If only one reservoir is studied, the material balance of the component x can be described by the following equations:

$$\frac{dm_x}{dt} = M_x - N_x \tag{5.2}$$

where m_x is the total amount of the component x in the reservoir and M_x and N_x are the rates of its introduction and sink, respectively.

The sink processes are usually described by the first-order equation $N_x = \lambda_x m_x$ where λ_x is the coefficient having the meaning of the mean sink rate and the dimensionality of y^{-1}. Then we have

$$\frac{dm_x}{dt} = M_x - \lambda_x m_x \tag{5.3}$$

When the quantity of the component is invariable, the equation of the steady state is obtained

$$M_x - \lambda_x m_x = 0 \tag{5.4}$$

The average residence time of the component x in the reservoir is determined by the equations

$$\bar{\tau}_x = m_x \cdot M_x^{-1} \quad (5.5), \qquad \bar{\tau}_x = \lambda_x^{-1} \tag{5.6}$$

The rates of the introduction and sink of the component, M_x and N_x are additive values. The introduction rate is determined by emission from natural and anthropogenic sources and for very many compounds also by the rate of their formation directly in the atmosphere as a result of chemical reactions. Sinks include the processes of transport into other basins (for example, from the troposphere into the stratosphere), the sorption and precipitation on the supporting surface and, finally, the processes of chemical transformation of the component x in the atmosphere.

The transport from one reservoir into another is characteristic of components with a relatively long residence time. The precipitation on the supporting surface is particularly important for low-volatile compounds forming a part of aerosol particles. The main sink of most volatile organic components is provided by chemical reactions in the atmosphere.

Figure 5.1 shows in simplified form the main types of chemical transformations occurring in the lower layers of the atmosphere. The scheme shows that

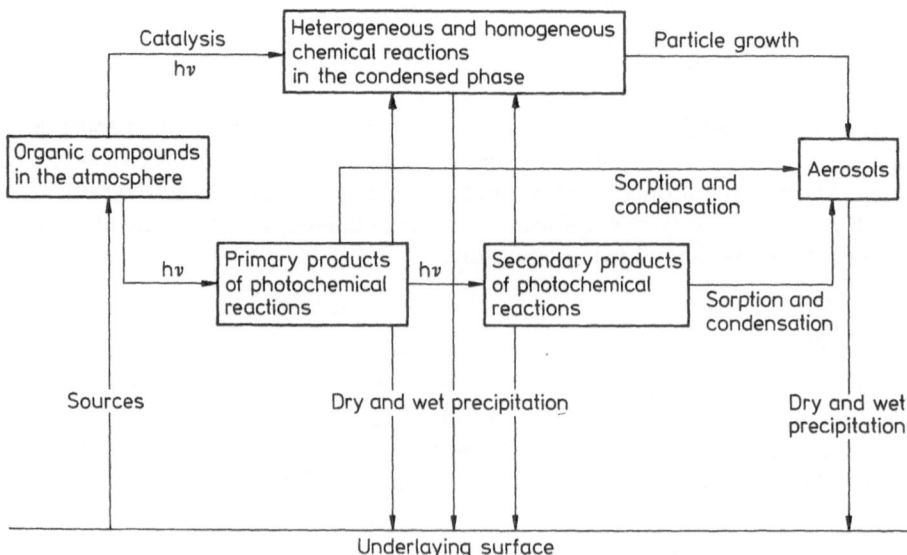

Fig. 5.1. Principal directions of transformation of volatile organic components in the atmosphere

electronically excited molecules of organic compounds participate in multi-stage (for simplicity, only two stages are shown in the scheme) gas-phase photochemical reactions. They serve on the one hand as sinks and on the other as sources of new compounds: the products of partial oxidation and decomposition of the initial components. Another pathway is the sorption of gaseous impurities by particles of solid and liquid aerosols and subsequent chemical transformations in the condensed phase. These reactions can proceed both in visible light and in the dark. A considerable role in the latter processes is evidently played by the catalysts which are probably mainly mineral components. In each of these stages, final inorganic oxidation products (carbon oxides and water) are formed and some intermediates can be partially removed from the atmosphere as a result of the processes of wet and dry precipitation. It is difficult to evaluate the significance of various pathways in the formation of sinks of individual organic compounds. It may only be assumed that heterogeneous processes are particularly important in the chemistry of the urban atmosphere containing large amounts of aerosols.

Since the beginning of the 1960s, the chemical sinks of minor gaseous atmospheric components have been intensively studied in many countries and at present some success in this respect has been achieved, in particular in the investigation of the gas-phase photochemical reactions. As to the transformations in the condensed phase, in this respect the investigations have been almost exclusively restricted to inorganic compounds: nitrogen and sulfur oxides, although many authors have emphasized the importance of these processes in the chemistry of organic components. This situation is probably caused by considerable experimental difficulties and the extreme complexity of the samples under investigation.

Considerable technical difficulties are also experienced in the investigation of the gas-phase photochemical reactions. At present, the experiments under natural conditions cannot be performed because of the complete absence of the required highly sensitive instrumentation. The study of the reaction kinetics is carried out by using the so-called smog chambers. They are reaction vessels the capacity of which is often many tens of cubic meters and the inner walls are made of inert materials. The light sources are lamps satisfactorily imitating the ultraviolet part of the solar spectrum at the Earth's surface and at various heights over it. The shortcomings of smog chambers result from the adsorption of components and other wall effects. The reagent concentrations in these chambers are usually many times higher than those characteristic even of the highly polluted urban atmosphere. However, the behavior of the irradiated mixtures in smog chambers serves as a starting point in the interpretation of the processes occurring in the open atmosphere.

The aim of the present chapter is to demonstrate the present level of understanding of the character of chemical processes in the atmosphere using a few examples. This chapter considers the main pathways of transformations of the principal classes of organic components of the atmosphere: alkanes, alkenes, arenes, carbonyl compounds and some other hydrocarbon derivatives as well as halogen-containing compounds in connection with the problems of the chemistry of the stratosphere.

Section 5.1 yields brief general information on some reactive particles and considers in general the atmospheric cycle of active oxygen and nitrogen compounds very important for the understanding of the subsequent sections.

5.1   Reactive Particles in the Atmosphere

Chemical transformations in the troposphere and the stratosphere are initiated mainly by the products of photolysis of such molecules as O_3, O_2, H_2O, N_2O and NO_2.

Ozone

The principal component determining the chemistry of the stratosphere is ozone. The main processes responsible for the ozone cycle in the stratosphere are described by the following reactions:

$$O_2 \xrightarrow{h\nu} O(^1D) + O(^3P) \qquad (\lambda < 175\,nm) \tag{5.7a}$$

$$\xrightarrow{} 2O(^3P) \qquad (\lambda < 242.2\,nm) \tag{5.7b}$$

$$O + O_2 + M \longrightarrow O_3 + M \tag{5.8}$$

$$O_3 + O \longrightarrow 2O_2 + 392\,kJ \tag{5.9}$$

$$O_3 \xrightarrow{h\nu} O_2(^1\Delta) + O(^1D) \qquad (\lambda < 310\,nm) \tag{5.10a}$$

The photodissociation coefficient J for the last reaction obtained by the integration over the entire ultraviolet and visible part of the solar spectrum is $10^{-2}s^{-1}$.

The release of thermal energy in ozone decomposition in Eqs. (5.9) and (5.10a) leads to the formation of an inversion layer at altitudes between 15 and 50 km. The maximum equilibrium ozone concentration is at an altitude of about 25 km. The above mechanism of photochemical formation and decomposition of ozone in the stratosphere is named the Chapman mechanism because he formulated it first in 1930.

The researchers explain the presence of ozone in the overland air layers by both the processes of its transport from the stratosphere and its formation as a result of a series of photochemical reactions with the participation of nitrogen oxides and organic substances. Ozone is photolytically dissociated by light in the visible region at a wavelength of less than 1180 nm. In this case molecular and atomic oxygen in the ground state are formed

$$O_3 \xrightarrow{h\nu} O_2(\Sigma_g^-) + O(^3P) \tag{5.10b}$$

A considerable effect of photochemical reactions on the content of O_3 in intermediate layers of the troposphere is confirmed by the results of the measurements carried out by Eastman J. A. and Stedman D. H. [313a]. During the solar eclipse in February 1979 its concentration at an altitude of 3.7 km decreased by 50%. The background concentrations of O_3 measured over the ocean surface in the Northern and Southern Hemispheres are at a level of 20–35 ppb, whereas in large cities and in the air of adjoining areas they attain 200–250 ppb.

Molecular and atomic oxygen

During the photolysis of ozone by light at wavelengths less than 310 nm, molecular oxygen is formed in the metastable excited state $O_2(^1\Delta)$ according to Eq. (5.10a). Its highest concentrations are observed at altitudes between 30 and 80 km with a maximum at 4×10^{10} cm^{-3} at an altitude of 50 km. The $O_2(^1\Delta)$ molecule is not very reactive. It reacts at the highest rate ($k = 4.4 \times 10^{15}$ cm^3 molecule^{-1}s^{-1}) with ozone:

$$O_2(^1\Delta) + O_3 \longrightarrow 2O_2 + O \tag{5.11}$$

The second particle formed in Eq. (5.10a), the metastable oxygen atom $O(^1D)$, is of much greater importance in the atmospheric chemistry. Its formation is also observed in the photodissociation of some other molecules:

$$N_2O \xrightarrow{h\nu} N_2 + O(^1D) \qquad (\lambda \leqslant 340 \text{ nm}) \tag{5.12}$$

$$NO_2 \xrightarrow{h\nu} NO + O(^1D) \qquad (\lambda \leqslant 244 \text{ nm}) \tag{5.13}$$

It is formed in the upper layers of the stratosphere together with the oxygen atom in the ground state when light at a wavelength less than 175 nm is absorbed according to Eq. (5.7). Metastable oxygen is an active particle with the lifetime of about 110 s. Its chemical and physical quenching takes place at a high rate. The rate constant for interaction with molecular oxygen

$$O(^1D) + O_2 \longrightarrow O(^3P) + O_2(^1\Sigma) \tag{5.14}$$

is 7×10^{-11} cm^3 molecule^{-1}s^{-1}. It is observed in considerable concentrations at altitudes of more than 20 km.

Atomic oxygen in the ground state is formed in the troposphere when ozone is dissociated by light at wavelengths less than 1180 nm (5.10b). It participates in the formation and decomposition of ozone in Eqs. (5.8) and (5.9).

Hydroxyl and hydroperoxide radicals

The hydroxyl radical, HO˙, is formed as a result of the photolysis of water and as an intermediate species in some reactions. Direct photodissociation of water evidently occurs only in the upper stratosphere and above the stratopause. The principal process of the formation of HO˙ in the stratosphere is the reaction of

H_2O with metastable oxygen:

$$O(^1D) + H_2O \longrightarrow 2HO^{\cdot} + 120.5\,kJ \tag{5.15}$$

An additional source is provided by the oxidation of methane and hydrogen:

$$O(^1D) + CH_4 \longrightarrow CH_3^{\cdot} + HO^{\cdot} \tag{5.16}$$

$$O(^1D) + H_2 \longrightarrow H^{\cdot} + HO^{\cdot} \tag{5.17}$$

These three processes have similar values of rate constants $(1–4) \times 10^{-10}\,cm^3$ molecule^{-1} s^{-1}.

In the troposphere, hydroxyl radical is also formed in decomposition reactions

$$HNO_2 \xrightarrow{\ h\nu\ } NO + HO^{\cdot} \qquad (\lambda < 400\,nm) \tag{5.18}$$

$$HNO_3 \xrightarrow{\ h\nu\ } NO_2 + HO^{\cdot} \qquad (\lambda \leqslant 335\,nm) \tag{5.19}$$

$$H_2O_2 \xrightarrow{\ h\nu\ } 2HO^{\cdot} \qquad (\lambda \leqslant 300\,nm) \tag{5.20}$$

The main sink of $HO^{\cdot}$ in the troposphere is considered to be its reactions with carbon oxide, organic compounds and nitrogen oxide

$$HO^{\cdot} + CO \longrightarrow CO_2 + H^{\cdot} \tag{5.21}$$

$$HO^{\cdot} + RH \longrightarrow R^{\cdot} + H_2O \tag{5.22}$$

$$HO^{\cdot} + NO + M \longrightarrow HONO + M \tag{5.23}$$

Hydroxyl radical is the only reactive intermediate species for which the concentration profile up to an altitude of 100 km has been determined experimentally. The results of aircraft surveys and land measurements [313b] show that its concentrations in the troposphere are at a level of $(0.5–0.4) \times 10^6$ and increase to $10 \times 10^6\,cm^{-3}$ in the stratosphere.

The oxidation of carbon monoxide to CO_2 is the final stage of decomposition of hydrocarbons and their derivatives in the atmosphere. Atomic hydrogen formed in Eq. (5.21) rapidly reacts with O_2 to give a hydroperoxide radical $HO_2^{\cdot}$:

$$H^{\cdot} + O_2 \longrightarrow HO_2^{\cdot} \tag{5.24}$$

It is also formed in the troposphere when O_3 and H_2O_2 are decomposed by the hydroxyl radical:

$$HO^{\cdot} + O_3 \longrightarrow HO_2^{\cdot} + O_2 \tag{5.25}$$

$$HO^{\cdot} + H_2O_2 \longrightarrow HO_2^{\cdot} + H_2O \tag{5.26}$$

It has been established that the hydroperoxide radical is an important intermediate in the processes of combustion and formation of photochemical smog. It actively participates in the oxidation of NO:

$$HO_2^{\cdot} + NO \longrightarrow NO_2 + HO^{\cdot} \tag{5.27}$$

Few available measurements indicate that the hydroperoxide radical is

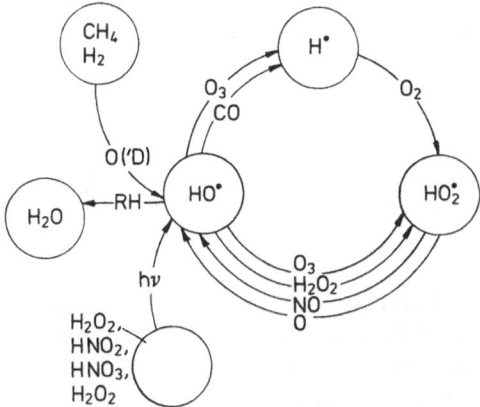

Fig. 5.2. Transformation of hydrogen-containing intermediates in the photochemical oxidation of methane in the troposphere

distributed more or less uniformly in the altitudes between 5 and 30 km at concentrations of 10^7–10^8 cm^{-3}.

The Nicolet scheme shown above (Fig 5.2) [313c] reflects the main relationships and pathways in the transformations of hydrogen-containing intermediates in the troposphere and stratosphere.

Nitrogen oxides

Nitrogen oxide (I) N_2O emitted almost exclusively by soil microorganisms is stable in the troposphere. Above the tropopause it undergoes photolysis according to Eqs. (5.12) and (5.28):

$$N_2O \xrightarrow{\ h\nu\ } O(^3P) + N_2 \qquad (5.28)$$

In the stratosphere, another sink for N_2O is the reaction with metastable atomic oxygen:

$$N_2O + O(^1D) \longrightarrow 2NO \qquad (5.29a)$$

$$N_2O + O(^1D) \longrightarrow N_2 + O_2 \qquad (5.29b)$$

The last two processes occur at approximately the same rates.

Nitrogen dioxide, NO_2, being irradiated with light at a wavelength of less than 244 nm, decomposes in the stratosphere to form $O(^1D)$ according to Eq. (5.13). The light at longer wavelengths (244–298 nm) gives NO and an oxygen atom in the ground state:

$$NO_2 \xrightarrow{\ h\nu\ } NO + O(^3P) \qquad (5.30)$$

This atom readily reacts with the oxygen molecule and as a result ozone is formed (5.8). Nigrogen oxide is again oxidized to NO_2 by the hydroperoxide radical with the evolution of HO$^{\bullet}$ (5.27). Atomic oxygen, ozone and the HO$^{\bullet}$ radical obtained in this cycle initiate the oxidation of hydrocarbons. These processes occur particularly actively in the strongly polluted atmosphere of cities in which automobile transport is the principal source of nitrogen oxides.

The cycle of nitrogen compounds in the troposphere is supplemented by the formation of nitrous acid (5.23), nitric acid and nitrogen trioxide:

$$NO_2 + HO^{\cdot} + M \longrightarrow HONO_2 + M \qquad (5.31)$$

$$NO_2 + O_3 \longrightarrow NO_3 + O_2 \qquad (5.32)$$

The decomposition of acids occurs according to Eqs.(5.18) and (5.19). The trioxide decomposes by the reaction with NO:

$$NO_3 + NO \longrightarrow 2NO_2 \qquad (5.33)$$

The background concentrations of NO and NO_2 over the oceans usually do not exceed 1 ppb. Single measurements of HNO_3 concentrations over continental non-urban areas in the West of the USA gave the values from 0.1 to 2.3 ppb [314]. Moreover, the total content of NO and NO_2 has always been higher than these values by the factor of not less than five. According to the data in ref. [315], the concentrations of HNO_2 in the air of Los Angeles in the summer months of 1980 were 4.5–8 ppb.

Sulfur dioxide

It is emitted into the atmosphere during the burning of fossil fuel, with volcanic gases and owing to the vital activity of soil microorganisms. The absorption of sunlight in the range of 340–400 nm leads to the appearance of photo-excited molecules, $SO_2(^3B_1)$, in the atmospheric air. The light absorbed in the range of 290–340 nm forms singlet excited molecules, $SO_2(^1A_1)$. The collision of these particles with the molecules of oxygen or nitrogen leads to the formation of triplet sulfur dioxide in the state $(^3B_1)$. Average SO_2 concentrations in the marine troposphere are on the level of 0.2 ppb, whereas in the air of the continental regions and in the cities they can exceed this value many tens of times [314].

The knowledge of rate constants of photochemical reactions being the principal sources and sinks for a given component, makes it possible to calculate its equilibrium concentration in the atmosphere. For example, if the main sources of atomic oxygen, $O(^3P)$, in the troposphere are the processes of the photolysis of ozone (5.10b) and nitrogen dioxide (5.30) and the main sink is the interaction with molecular oxygen, then the equilibrium concentration will be determined by the equation:

$$[O(^3P)] = \frac{J_{(5.10b)}[O_3] + J_{(5.30)}[NO_2]}{k_{(5.8)}[O_2] \cdot [M]} \qquad (5.35)$$

5.2   Atmospheric Chemistry of Hydrocarbons

Alkanes

Oxidative decomposition of saturated hydrocarbons starts with the abstraction of a hydrogen atom in the interaction with a hydroxyl radical. The alkyl radical formed in the first stage attaches an oxygen molecule to give a new unstable particle, an alkylperoxide radical:

$$R-H + HO^{\cdot} \longrightarrow R^{\cdot} + H_2O \tag{5.36}$$

$$R^{\cdot} + O_2 + M \longrightarrow ROO^{\cdot} + M \tag{5.37}$$

Further transformations may occur in several alternative directions with the participation of NO and NO_2 molecules or radical particles:

$$RO_2^{\cdot} + NO \longrightarrow RO^{\cdot} + NO_2 \tag{5.38}$$

$$RO_2^{\cdot} + NO + M \longrightarrow RONO_2 + M \tag{5.39}$$

$$RO_2^{\cdot} + NO_2 + M \longrightarrow ROONO_2 + M \tag{5.40}$$

$$RO_2^{\cdot} + HO_2^{\cdot} \longrightarrow ROOH + O_2 \tag{5.41}$$

In the first of these reactions, a new alkoxy radical is formed. Among the other reactions only one Eq. (5.39) yields relatively stable product, alkyl nitrate, whereas peroxide compounds formed in the reaction with NO_2 and a hydroperoxide radical readily decompose as a result of the action of light or the attack of a hydroxyl:

$$ROOH \xrightarrow{h\nu} RO^{\cdot} + HO^{\cdot} \tag{5.42}$$

$$ROOH + HO^{\cdot} \longrightarrow RO_2^{\cdot} + H_2O \tag{5.43}$$

Hence, only the formation of alkyl nitrates is a reaction competing with (5.38).

It is assumed that this reaction proceeds via the three-membered transition state according to the following scheme:

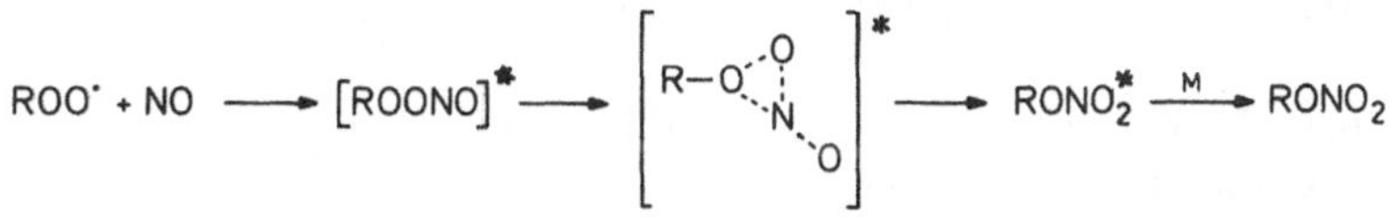

$$ROO^{\cdot} + NO \longrightarrow [ROONO]^* \longrightarrow \left[R-O \underset{N}{\overset{O}{\cdots}} \underset{O}{\cdots} \right]^* \longrightarrow RONO_2^* \xrightarrow{M} RONO_2 \tag{5.44}$$

where the molecule M accepts the excess energy of the excited particles. It has been established [316] that the formation of alkyl nitrates becomes really a competing direction with respect to the process (5.38) only when long-chain alkanes are oxidized: for ethane oxidation, the yield of ethyl nitrate is about 1% based on the reacting hydrocarbon, whereas in the case of n-octane the yield attains 33%.

Alkoxy radicals can subsequently be transformed in several directions. The first of them leads to the formation of a carbonyl compound as a result of the abstraction of a hydrogen atom from the α-position with respect to the radical center after a collision with oxygen:

$$RCH(O^{\cdot})R' + O_2 \longrightarrow R\overset{O}{\overset{\|}{C}}R' + HO_2^{\cdot} \tag{5.45}$$

The second direction, the cleavage of the radical, also gives a carbonyl compound but with fewer carbon atoms than in the initial hydrocarbon:

$$RCH(O^{\cdot})R' \longrightarrow RCH{=}O \ + \ R'^{\cdot} \tag{5.46}$$

The third direction is the intramolecular shift of a hydrogen atom. This shift can occur only for alkanes with a long chain of carbon atoms (not less than four) because it proceeds via the six-membered transition state:

$$\tag{5.47}$$

Hence, alkoxy radicals, $RO^{\cdot}$, are the precursors of carbonyl compounds (Eqs. 5.45 and 5.46) and new alkyl radicals (Eqs. 5.46 and 5.47). Carbonyls undergo subsequent photolytic decomposition:

$$R-\overset{\overset{\text{O}}{\|}}{C}-R \xrightarrow{\ h\nu\ } R^{\cdot} + R-\overset{\cdot}{C}{=}O \tag{5.48}$$

and the resulting alkyl radicals are involved in the processes already described with the participation of oxygen molecules and nitrogen oxide (5.37 and 5.38). Very important intermediate particles, acetyl radicals $R\overset{\cdot}{C}{=}O$, are formed in the photolysis of carbonyl compounds. They rapidly react with an oxygen molecule and nitrogen dioxide to give relatively stable peroxyacyl nitrates which are very harmful to man:

$$R-\overset{\cdot}{C}{=}O + O_2 \longrightarrow R-C\overset{\nearrow O}{\underset{\searrow OO^{\cdot}}{}} \tag{5.49}$$

$$R-C\overset{\nearrow O}{\underset{\searrow OO^{\cdot}}{}} + NO_2 \longrightarrow R-C\overset{\nearrow O}{\underset{\searrow OONO_2}{}} \tag{5.50}$$

The best known representative of this class of organic compounds is peroxyacetyl nitrate, (see below).

The main directions of transformations of the most abundant organic component of the atmosphere, methane, are shown in Fig. 5.3. It can be seen from this scheme that the final and relatively long-lived product (its lifetime in the troposphere is about three months) is carbon monoxide. At present, it is recognized that the photochemical oxidation of methane and other hydrocarbons of natural origin is the main source of atmospheric CO, and its total emission is evaluated to be from 3×10^{14} to $50 \times 10^{14} \, g \, y^{-1}$.

Further oxidation of CO proceeds according to the following scheme:

$$CO + HO^{\cdot} \longrightarrow H^{\cdot} + CO_2 \tag{5.51}$$

$$H^{\cdot} + O_2 + M \longrightarrow HO_2^{\cdot} + M \tag{5.52}$$

$$HO_2^{\cdot} + NO \longrightarrow NO_2 + HO^{\cdot} \tag{5.27}$$

$$\overline{CO + O_2 + NO \longrightarrow NO_2 + CO_2} \tag{5.53}$$

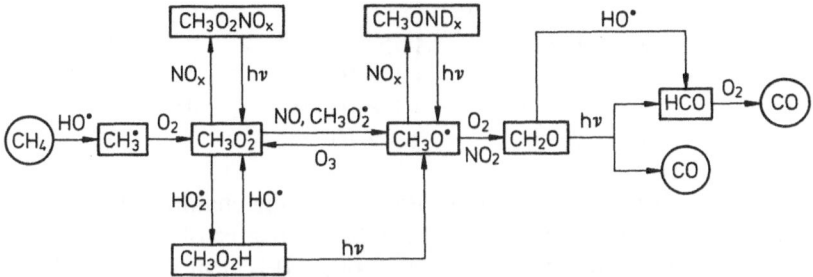

Fig. 5.3. Main trends of the transformation of intermediates in the photochemical oxidation of methane in the troposphere

The total result of CO oxidation (Eq. 5.53) is the formation of NO_2 which is involved in the processes leading to ozone (Eqs. 5.30 and 5.8).

The rate of reaction with the hydroxyl radical strongly depends on the structure of alkane molecules. The secondary and tertiary hydrogen atoms of methylene ($-CH_2-$) and methyne ($-CH\diagdown$) groups are more readily abstracted than the primary atoms. It is clear, therefore, that the rate of reaction (5.36) should increase on passing from methane to its homologues and from linear alkanes to their branched isomers. This is confirmed by the values of rate constants for reactions of the alkane series with a hydroxyl at 298 K (in cm^3 molecule^{-1} s^{-1}) according to the data in refs. [317, 318].

Methane	8.0×10^{-15}	Butane	2.6×10^{-12}
Ethane	2.7×10^{-13}	Pentane	5.0×10^{-12}
Propane	1.2×10^{-12}	2,3-Dimethylbutane	6.3×10^{-12}

An empirical equation has been proposed [318] for calculating the rate constant for the reaction of isomeric alkanes with a hydroxyl radical:

$$k = (1.01 \times 10^{-12})N_p e^{-823/T} + (2.41 \times 10^{-12})N_s e^{-428/T} + (2.10 \times 10^{-12})N_t,$$

where N_p, N_s and N_t are the numbers of primary, secondary and tertiary hydrogen atoms in the alkane molecule, respectively, and k is expressed in cm^3 molecule^{-1} s^{-1}. This equation allows one to calculate the rate constant for alkanes containing not less than three carbon atoms in the molecule with a precision of not less than $\pm 20\%$ in the temperature range 300–500 K.

Alkenes

Hydrocarbons containing a double carbon–carbon bond are highly reactive. In this case under the conditions of the troposphere, the processes of attachment on the double bond play the major role. The rate constants (in cm^3 molecule^{-1} s^{-1}) for the reaction of lower alkenes with some particles [318] are given below.

	HO˙	O_3	$O(^3P)$
Ethylene	8.1×10^{-12}	1.75×10^{-18}	7.3×10^{-13}
Propylene	2.5×10^{-11}	1.1×10^{-17}	4.0×10^{-12}
1-Butene	3.3×10^{-11}	1.1×10^{-17}	4.2×10^{-12}
trans-2-Butene	7.0×10^{-11}	2.0×10^{-16}	2.3×10^{-11}

The study of the behavior of various alkenes in smog chambers has shown that the rate of hydroxyl addition increases with the degree of substitution of hydrogen atoms at the double bond by alkyl groups. It is clear from the above data that on passing from a hydrocarbon with an unsubstituted double bond (ethylene) to that in which it is substituted by two methyl groups (2-butene), the rate of addition of the HO˙ radical increases by almost one order of magnitude. Alkenes with tetrasubstituted double bonds are found to be the most reactive.

The addition of a hydroxyl radical initiates a chain of transformations the sequence of which may be followed taking propylene as an example:

$$CH_3CH{=}CH_2 + HO^{\cdot} \longrightarrow CH_3\dot{C}HCH_2OH \qquad (5.54)$$

$$\longrightarrow CH_3CH(OH)\dot{C}H_2 \qquad (5.55)$$

According to the data in ref. [318], the reaction takes place mainly according to Eq. (5.54). Further transformations of hydroxyalkyl radicals proceed according to the scheme shown above for alkyl radicals and involving the attachment of an oxygen molecule, the oxidation of nitrogen oxide and fragmentation:

$$CH_3\dot{C}HCH_2OH + O_2 \longrightarrow CH_3CH(OO^{\cdot})CH_2OH \qquad (5.56)$$

$$CH_3CH(OO^{\cdot})CH_2OH + NO \longrightarrow CH_3CH(O^{\cdot})CH_2OH + NO_2 \qquad (5.57)$$

$$CH_3CH(O^{\cdot})CH_2OH + O_2 \longrightarrow CH_3COCH_2OH + HO_2^{\cdot} \qquad (5.58)$$

$$CH_3CH(O^{\cdot})CH_2CH \longrightarrow CH_3CH{=}O + \dot{C}H_2OH \qquad (5.59)$$

$$\dot{C}H_2OH + O_2 \longrightarrow CH_2{=}O + HO_2^{\cdot} \qquad (5.60)$$

Formic acid (HCOOH) is one of the main products of propylene oxidation in smog chambers in the presence of NO. It can be formed as a result of successive reactions:

$$\dot{C}H_2OH + O_2 \longrightarrow \dot{O}OCH_2OH \qquad (5.61)$$

$$\dot{O}OCH_2OH + NO \longrightarrow \dot{O}CH_2OH + NO_2 \qquad (5.62)$$

$$\dot{O}CH_2OH + O_2 \longrightarrow HCOOH + HO_2^{\cdot} \qquad (5.63)$$

It can be clearly seen that the photochemical oxidation of alkenes under the conditions of the troposphere considerably affects ozone formation. Even incomplete oxidation of the propylene molecule leads to the formation of four NO_2 molecules:

$$C_3H_6 + 4O_2 + 4NO \longrightarrow CH_3CH{=}O + 4NO_2 + CO_2 + H_2O \qquad (5.64)$$

A part of the resulting NO_2 is involved in a reaction with acetaldehyde leading to peroxyacetyl nitrate:

$$CH_3CH{=}O \xrightarrow[HO^\cdot]{h\nu} CH_3\dot{C}{=}O \xrightarrow{O_2} CH_3C\overset{\displaystyle{\nearrow}O}{\underset{OO^\cdot}{\searrow}} \xrightarrow{NO_2} CH_3C\overset{\displaystyle{\nearrow}O}{\underset{OONO_2}{\searrow}} \quad (5.65)$$

In the general case, the accumulation of ozone as a result of interaction of alkenes with a hydroxyl in the presence of nitrogen oxide depends on the ratio of the reagents. Akimoto H. and Sakamaki F. [319] have established that the rate of ozone formation is approximately proportional to the product of initial concentrations of propylene and $HO^\cdot$ radicals under the condition $[C_3H_6]/[NO_x]_0 \geqslant 5$. In this case the effective rate constant for ozone formation is high: it is $4.1 \times 10^{-11}\,cm^3\,molecule^{-1}\,s^{-1}$. However, these ratios of alkene and nitrogen oxide concentrations are typical only of the urban atmosphere. In the air of rural and background areas, total alkene concentrations are low, and under these conditions they serve rather as a sink than a source for ozone.

If the above rates of addition of hydroxyl and ozone to alkenes are compared, a wrong impression may be formed that the reaction with ozone is not competitive: its rate is lower by 5–6 orders of magnitude than that of the addition of the hydroxyl radical. However, the concentration of ozone even in the background areas is higher by about 6–7 orders of magnitude than that of hydroxyl radicals. Hence, the reaction with ozone evidently serves as one of the main tropospheric sinks of alkenes.

In the first stage, ozonides are formed; they decompose to biradicals:

$$CH_3-CH{=}CH_2 \;+\; O_3 \longrightarrow CH_3-\underset{\overset{|}{O}}{CH}-\underset{\overset{|}{O}}{CH_2} \Big/\!\!\!\Big\backslash$$

(5.66)

$$CH_3-\underset{}{CH}-CH_2 \longrightarrow \begin{array}{c} H_2\dot{C}OO \\ + \\ CH_3CH{=}O \end{array}$$

$$CH_3-CH-CH_2 \longrightarrow \begin{array}{c} CH_3\dot{C}HOO^\cdot \\ + \\ H_2C{=}O \end{array}$$

It is assumed that biradicals subsequently decompose according to the following scheme including intermediates with the dioxirane structure:

$$\dot{C}H_2OO^\cdot \longrightarrow \left[H_2C\overset{\overset{O}{\diagup}}{\underset{O}{\diagdown}} \right]^* \begin{array}{l} \longrightarrow 2H^\cdot + CO_2 \\ \longrightarrow H_2O + CO \\ \longrightarrow H_2 + CO_2 \\ \xrightarrow{M} HCOOH \end{array} \qquad (5.67)$$

$$CH_3\dot{C}HCOO^\cdot \longrightarrow \left[CH_3-CH\overset{\overset{O}{\diagup}}{\underset{O}{\diagdown}} \right]^* \begin{array}{l} \longrightarrow CH_3^\cdot + CO_2 + H^\cdot \\ \longrightarrow CH_3^\cdot + CO + HO^\cdot \\ \xrightarrow{M} CH_3COOH \end{array}$$

$$\Big\downarrow$$

(5.68)

$$CH_3-\underset{\overset{\|}{O}}{OCH} \xleftarrow{M} \left[CH_3-OCH{=}O \right]^* \begin{array}{l} \longrightarrow CH_4 + CO_2 \\ \longrightarrow CH_3O^\cdot + CO + H^\cdot \\ \longrightarrow CH_3^\cdot + CO_2 + H^\cdot \end{array}$$

Another direction of biradical transformation may be the oxidation of nitrogen oxide:

$$\dot{C}H_2O\dot{O} + NO \longrightarrow CH_2=O + NO_2 \tag{5.69}$$

$$CH_3\dot{C}HOO^\cdot + NO \longrightarrow CH_3CH=O + NO_2 \tag{5.70}$$

The photochemistry of dienes, cycloalkenes and cycloalkadienes which include isoprene and most monoterpene hydrocarbons· detected in the open atmosphere [149, 320–325] are of great scientific interest.

The addition of a hydroxyl radical to the isoprene molecule takes place at a rate three times that of a similar reaction with propylene (k = 9.3 × 10^{-11} cm^3 molecule^{-1} s^{-1}). The main reaction products are methyl vinyl ketone, methyl acrolein and formaldehyde.

$$\tag{5.71}$$

The $\dot{C}H_2OH$ radical is a formaldehyde precursor (see Eq. 5.60). All these products readily undergo photolysis with the formation of new radical particles. The fact that although the emission rate of isoprene by vegetation is high it is not accumulated even in the atmosphere of forests is probably associated with its high reactivity. According to the opinion of some authors, the processes of isoprene oxidation provide over 50% of the total amount of peroxide radicals in the air of rural areas.

The addition of ozone to the isoprene molecule takes place on one of the two double bonds. The first stages of the process are shown in scheme (5.72).

$$\tag{5.72}$$

Carbon oxides are the final reaction products. There are reasons for assuming that isoprene oxidation is one of the principal CO sources in the troposphere [320].

Monoterpene hydrocarbons are even more active in photochemical reactions. The oxidation of these unsaturated compounds is initiated by the addition of ozone, a hydroxyl radical and an NO_3 radical. The NO_3 radical plays an important role in chemical transformations proceedings in the troposphere at night time. Table 5.1 gives the rate constants for the reactions of various terpenes

Table 5.1. Rate constants and lifetimes of monoterpenes in reactions with ozone and radicals [324]

Monoterpene	k, cm^3 molecule^{-1} s^{-1}			Lifetime[a] in reactions with		
	O_3	OH	NO_3	O_3 (h)	OH (h)	NO_3 (min)
β-Pinene	2.1×10^{-17}	7.8×10^{-11}	2.5×10^{-12}	18	3.6	28
α-Pinene	8.4×10^{-17}	6.0×10^{-11}	6.1×10^{-12}	4.6	4.6	11
3-Carene	1.2×10^{-16}	8.7×10^{-11}	1.1×10^{-11}	3.2	3.2	6.3
Limonene	6.4×10^{-16}	1.4×10^{-10}	1.4×10^{-11}	0.6	2.0	5.0
Myrcene	1.2×10^{-15}	1.9×10^{-10}	1.1×10^{-11}	0.3	1.5	6.3
cis-Ocymene	2.0×10^{-15}	2.3×10^{-10}	2.4×10^{-11}	0.2	1.2	2.9
α-Phellandrene	1.2×10^{-14}	1.7×10^{-10}	9.1×10^{-11}	0.03	1.6	0.7
β-Phellandrene	1.8×10^{-16}	1.4×10^{-10}	1.3×10^{-11}	2.1	2.0	5.3
α-Terpinene	8.8×10^{-14}	2.1×10^{-10}	1.9×10^{-10}	0.004	1.3	0.36
Terpinolene	1.0×10^{-14}	2.0×10^{-10}	7.1×10^{-11}	0.04	1.4	1.0

[a]Calculated for the concentrations of O_3-30 ppb, HO-1×10^6 cm^{-3}, NO_3-2.4×10^8 cm^{-3} (at night time)

with these particles. The same table gives the calculated lifetime of monoterpenes in the over ground layer of the atmosphere at fixed concentrations of active particles. It is clear that the differences in the reactivities of individual $C_{10}H_{16}$ hydrocarbons are very great: the rate constants for the addition of ozone and NO_3 to β-pinene and α-terpinene differ by several orders of magnitude.

Experiments in smog chambers [320] show that only 5 min after the mixing of 1.7 ppm of α-pinene and 12 ppm of ozone in the dark the entire terpene has disappeared. The IR spectra of the gas mixture exhibit the presence of new compounds with absorption bands characteristic of the carbonyl group. Subsequent irradiation of the reaction mixture led to the appearance of CO and CO_2 the yield of which attained 30% of the theoretical based on hydrocarbon.

The relatively low yield of gaseous compounds is evidently due to the formation of aerosol particles by most products. In fact, the authors of ref. [321] have detected in aerosols more than 50% of oxidation products of limonene in the presence of nitrogen oxide in the smog chamber. In the dark reaction of limonene and terpinolene with ozone, conversion of products into aerosols attained almost 100%. These aerosols consisted of the products of oxidation and the ring opening of polycarbonyl compounds, aldo- and ketoacids, etc.

The pressure of the saturated vapor of initial terpene hydrocarbons at room temperature is on the level of 1 mm Hg and is lower by 4–5 orders of magnitude for the compounds formed by oxidation. These compounds are probably readily involved in the processes of aerosol formation (nucleation). The investigations carried out in recent years in the forests of the tropical zone [326], in south-east France [327], in Siberia [328] and Lithuania [329] have confirmed the suggestion of Went F. [236] that the processes of photochemical oxidation of biogenic organic compounds are an important source of atmospheric aerosols.

The investigations in coniferous forests of Lithuania have shown that in the period of blue haze formation, the concentration of particles with a diameter of up to $0.3\,\mu m$ increases rapidly. The volume rate of this increase was $0.48\,\mu m^3\,cm^{-3}\,h^{-1}$. The study of thermal stability of this aerosol shows that volatile organic components with the evaporation temperature of up to $120\,^{\circ}C$ predominate in it (60%). About 30% consisted of higher boiling point organic compounds and sulfates [329]. The scale of aerosol formation with the participation of biogenic components is shown by the evaluation of the authors of ref. [326] according to which the flow of C_{org} in aerosols from the forests of the tropical zone alone is $20 \times 10^{12}\,gC\,y^{-1}$.

The contribution of photochemical oxidation of terpenes to the atmospheric reservoir of carbon oxides remains uncertain because so far the problems of the yield of gaseous reaction products and the further fate of aerosol particles containing relatively high boiling compounds have not been solved. There are reasons for supposing that organic aerosol constituents undergo further extensive decomposition as a result of photocatalytic processes (see Sect. 5.6).

Aromatic hydrocarbons

Benzene and its nearest homologues make up 30–40% of the total amount of hydrocarbons in urban air. Their fraction in the organic component of the atmosphere of background areas is much smaller. The reason for this is the predominant anthropogenic character of aromatic hydrocarbons emitted into the atmosphere mainly with the exhaust gases of automobile transport. Some natural sources of aromatic hydrocarbons have also been found (e.g. they are formed during forest fires and are emitted by plants and volcanoes) but their contribution to the total emission is probably small.

The reactivity of aromatic hydrocarbons is profoundly affected by the degree of substitution of the benzene ring and increases with the number of alkyl substituents. Several directions of arene transformations have been indicated [318, 330–333]. The main ones are the substitution on the benzene ring and the processes of ring opening with subsequent fragmentation. As a result of substitution reactions, benzene and toluene form nitro- and hydroxy compounds, such as nitrobenzene, nitrotoluenes, nitrophenols and cresols. Many of them have been detected not only in smog chambers but also in urban air. In particular, seven nitrophenols have been found in the atmosphere of Yokohama [333].

The yields of the products of substitution on the aromatic ring do not amount to more than 15% based on the reacting hydrocarbon. The main transformation trend providing over 80% of the products is ring opening. This process starts with the addition of a hydroxyl radical and the formation of an adduct radical followed by that of a peroxide radical

$$(5.73)$$

In order to interpret the formation of reaction products (α-dicarbonyl compounds), Atkinson R. et al. have proposed a mechanism involving intra-molecular cyclization of the peroxide radical and subsequent interaction of the bicyclic intermediate with O_2 and NO

$$(5.74)$$

The fragmentation of the resulting alkoxyl radical yields dicarbonyl com-pounds. In the case of benzene, they are glyoxal and butenedial:

$$O=CH-CH=O \ + \ O=CH-CH=CH-CH=O \qquad (5.75)$$

The photochemical oxidation of benzene homologues leads to a great number of various products. In the case of toluene, 27 components have been detected including 4 mono- and 10 dicarbonyl compounds, the products of ring opening. The other components were formed as a result of substitution reactions on the aromatic ring (nitrotoluenes, nitrocresols, benzaldehyde and its derivatives). The principal reaction products, dicarbonyl compounds, readily undergo photodis-sociation to radicals. The rate of photolytic dissociation of methylglyoxal, one of the main products of toluene oxidation, exceeds that of formaldehyde by a factor of over 15.

The low vapor pressure of many products of photooxidation of aromatic hydrocarbons favors their participation in the processes of aerosol formation [334]. Free radicals formed in the photolysis of carbonyl compounds may evidently also serve as condensation nuclei [330].

5.3   Oxygen-, Nitrogen- and Sulfur-Containing Compounds

Hydrocarbon derivatives make up a considerable fraction of minor organic gas components of the atmosphere. As has been shown in the preceding section, many of them (carbonyl compounds, acids and alcohols) are not only emitted by land sources but also formed directly in the troposphere as a result of hydrocarbon oxidation.

Aldehydes and ketones

Carbonyl compounds are characterized by two directions of transformations: photolytic dissociation and oxidation initiated by radicals, mainly by the hydroxyl radical.

The photolytic dissociation of carbonyl compounds proceeds according to Eq. (5.48). It should be noted that recent investigations have established a rather complex character of aldehyde photooxidation. In particular, it has been shown that the quantum yields of H_2 and CO in the photolysis of formaldehyde in the oxygen-lean atmosphere may attain 7.07 and 11.6, respectively. This fact suggested that the reaction follows a chain mechanism involving the formation of short-lived adducts of formaldehyde with the $HO^{\cdot}$ and $HO_2^{\cdot}$ radicals [355].

The photooxidation of aldehydes initiated by the $HO^{\cdot}$ radical starts from the abstraction of a hydrogen atom from the carbonyl group as is shown taking acetaldehyde as an example:

$$CH_3CH{=}O + HO^{\cdot} \longrightarrow CH_3\dot{C}O + H_2O \quad (k = 1.2 \times 10^{-11}\,cm^3 \quad (5.76)$$

$$molecule^{-1}\,s^{-1})$$

$$CH_3\dot{C}0 + O_2 \longrightarrow CH_3C(O)OO^{\cdot} \tag{5.49}$$

$$CH_3C(O)OO^{\cdot} + NO \longrightarrow CH_3C(O)O^{\cdot} + NO_2 \tag{5.77}$$

The acetyl radical interacting with an oxygen molecule abstracts CO_2:

$$CH_3C(O)O^{\cdot} + O_2 \longrightarrow CH_3OO^{\cdot} + CO_2 \tag{5.78}$$

Taking into account further transformations of the methyl peroxide radical shown in Fig. 5.3, one can write a general equation for the complete oxidation of acetaldehyde:

$$CH_3CH{=}O + O_2 + 4NO \xrightarrow{h\nu} 2CO + 4NO_2 + H_2O + 2HO^{\cdot} \tag{5.79}$$

It can be seen from Eq. (5.79) that the atmospheric chemistry of aliphatic aldehydes profoundly affects the cycle of nitrogen oxides: the oxidation of one acetaldehyde molecule is accompanied by that of four NO molecules.

The photochemical oxidation of ketones starts with the abstraction of a hydrogen atom by a hydroxyl radical. The rate constants for the first members of the homologous series of aliphatic ketones, acetone and 2-butanone, are 2.8 $\times 10^{-13}$ and $8.8 \times 10^{-13}\,cm^3$ molecule$^{-1}\,s^{-1}$, respectively, but for 4-methyl-2-pentanone, the tertiary hydrogen atom is much more readily abstracted ($k = 1.3 \times 10^{-11}\,cm^3$ molecule$^{-1}\,s^{-1}$) [316]. The first oxidation stages for 2-butanone are described by the following elementary reactions:

$$CH_3CH_2COCH_3 + HO^{\cdot} \longrightarrow CH_3\dot{C}HCOCH_3 + H_2O \tag{5.80}$$

$$CH_3\dot{C}HCOCH_3 + O_2 \longrightarrow CH_3CH(OO^{\cdot})COCH_3 \tag{5.82}$$

$$CH_3CH(O^{\cdot})COCH_3 \longrightarrow CH_3CH{=}O + CH_3\dot{C}O \tag{5.83}$$

The study of acetone photolysis at air pressures varying from 20 to 745 Torr shows that in the lower layers of the troposphere the photolytic dissociation of ketones is of limited importance. The reaction yield of ketone photo-dissociation decreases with increasing pressure tending to a constant value, which shows that even at an air pressure of 1 atm about 7.7% of photo-excited acetone molecules undergo photolytic dissociation [336].

Carbonic acids and alcohols

The photochemistry of carbonic acids has not been investigated in detail. Little information is available in the literature about the products of their oxidation and photolysis in the atmosphere. However, these data are of interest because lower carbonic acids are an important intermediate in alkene sinks (see Eqs. 5.63, 5.67 and 5.68).

The photolysis of formic acid in the near UV range yields the following radicals:

$$HCOOH \xrightarrow{h\nu} H^{\cdot} + \dot{C}OOH \tag{5.84a}$$

$$\xrightarrow{h\nu} HO^{\cdot} + H\dot{C}O \tag{5.84b}$$

$$\xrightarrow{h\nu} H^{\cdot} + HCOO^{\cdot} \tag{5.84c}$$

It is assumed that the fragmentation of the molecule with the formation of stable products:

$$HCOOH \xrightarrow{h\nu} CO_2 + H_2 \tag{5.85}$$

does not play a significant role. The photodissociation coefficients and the quantum yields of the products of these elementary reactions are unknown.

The principal photochemical sinks for carboxylic acids in the troposphere are probably the reactions with the $HO^{\cdot}$ radical. The rate constant $(1 \times 10^{13} \, cm^3 \, molecule^{-1} s^{-1})$ for the reaction of a hydroxyl with formic, acetic, propionic and *n*-butyric acids are 3.2 ± 0.2, 6.0 ± 0.8, 16.0 ± 1.2 and 18.0 ± 1.6, respectively [337]. The rate of the process is of the same order of magnitude as that of hydrogen abstraction from *n*-alkanes (see Sect. 5.2). This fact suggests that the oxidation of formic acid homologues starts from an attack of the hydroxyl at the hydrogen of the alkyl group:

$$RCH_2COOH + HO^{\cdot} \longrightarrow R\dot{C}HCOOH + H_2O \tag{5.86}$$

The photochemical reactions of alcohols have been studied to a still lesser extent. The rate constants for the reaction of methanol and ethanol with an $HO^{\cdot}$ radical are 1.10×10^{-12} and $3.3 \times 10^{-2} \, cm^3 \, molecule^{-1} s^{-1}$ [317, 338]. It may be assumed that the $HO^{\cdot}$ radical abstracts a hydrogen atom from both the alkyl and the hydroxyl group of the alcohol:

$$CH_3OH + HO^{\cdot} \longrightarrow \dot{C}H_2OH + H_2O \tag{5.87a}$$

$$\longrightarrow CH_3O^{\cdot} + H_2O \tag{5.87b}$$

The investigation of the photochemical behavior of the lower alcohols is of great importance because they are widely used as automobile fuel. The authors of ref. [339] have established that when CH_3OH and NO_2 simultaneously emitted into the urban atmosphere with the exhaust gases of motor cars react with each other, methyl nitrite is formed. They have proposed a mechanism according to which in the first stage an asymmetric dimer NO_2 is formed reacting bimolecularly with CH_3OH.

$$2NO_2 + M \rightleftharpoons N_2O_4 + M \qquad (5.88)$$

$$N_2O_4 + CH_3OH \longrightarrow CH_3ONO + HNO_3 \qquad (5.89)$$

Amines

Much attention has been devoted to the behavior of aliphatic amines in gas-phase photochemical and dark reactions because of the problem of formation of carcinogenic N-nitrosoamines. It has been established that in the visible light alkylamines are attacked by the $HO^\cdot$ radical at a high rate. The rate constants for the reaction of methyl- and ethylamine are 2.2×10^{-11} and $2.8 \times 10^{-11} \, cm^3$ molecule$^{-1} s^{-1}$ [338]. The oxidation of triethylamine begins with the abstraction of a hydrogen atom in the α-position with respect to the nitrogen atom:

$$(C_2H_5)_2NCH_2CH_3 + HO^\cdot \longrightarrow (C_2H_5)_2N\overset{\cdot}{C}HCH_3 + H_2O \qquad (5.90)$$

The next stages are similar to those postulated for the transformations of alkyl radicals: a peroxide radical is formed in the reaction with O_2 followed by the oxidation of NO to NO_2 and the formation of the amine-alkoxide radical. This radical undergoes fragmentation with the breaking of the C—C or C—N bond:

$$(C_2H_5)_2NCH(O^\cdot)CH_3 \longrightarrow CH_3CH{=}O + (C_2H_5)_2N^\cdot \qquad (5.91a)$$

$$\longrightarrow (C_2H_5)_2NCH{=}O + CH_3^\cdot \qquad (5.91b)$$

or loses a hydrogen atom in a collision with the oxygen molecule:

$$(C_2H_5)_2NCH(O^\cdot)CH_3 + O_2 \longrightarrow (C_2H_5)_2NCOCH_3 + HO_2^\cdot \qquad (5.92)$$

The radical formed in Eq. (5.91a) reacts with nitrogen oxides to give diethylnitrosoamine and diethylnitroamine

$$(C_2H_5)N^\cdot + NO \longrightarrow (C_2H_5)_2NNO \qquad (5.93a)$$

$$(C_2H_5)_2N^\cdot + NO_2 \longrightarrow (C_2H_5)_2NNO_2 \qquad (5.93b)$$

Apart from the products shown in the schemes, large amounts of PAN are accumulated in the reaction medium.

Sulfur-containing compounds

The oxidation of dimethyl sulfide which is the principal organic sulfur compound of natural origin, may be represented by the following scheme:

$$CH_3SCH_3 \xrightarrow[-H_2O]{+HO} CH_3S\overset{\cdot}{C}H_2 \xrightarrow{+O_2} CH_3SCH_2OO^\cdot \xrightarrow[-NO_2]{+NO}$$

$$CH_3SCH_2O^\cdot \longrightarrow CH_2O + CH_3\overset{\cdot}{S} \qquad (5.94)$$

Further transformations of the $CH_3S^\cdot$ radical lead to several stable sulfur

compounds invariably found in experiments in smog chambers. They are mainly
methane sulfonic acid and sulfur dioxide. The formation of the acid may proceed
in two ways:

$$CH_3S^. \xrightarrow{HO^.} CH_3SOH \xrightarrow{O_2} CH_3SO_2OH \tag{5.95}$$

$$CH_3S^. \xrightarrow{O_2} CH_3SO_2^. \xrightarrow{HO} CH_3SO_2OH \tag{5.96}$$

Sulfur dioxide may also be formed during various reactions:

$$CH_3\dot{S} + O_2 \longrightarrow CH_3SO_2^. \longrightarrow CH_3^. + SO_2 \tag{5.97}$$

$$CH_3SO_2^. + NO \longrightarrow CH_3SO^. + NO_2 \tag{5.98}$$

$$CH_3SO^. \longrightarrow CH_3^. + SO \tag{5.99}$$

$$SO + O_3 \longrightarrow SO_2 + HO_2^. \tag{5.100}$$

$$SO + NO_2 \longrightarrow SO_2 + NO \tag{5.101}$$

The formation of methane sulfonic acid may be due to an alternative
mechanism involving in the intermediate stage the addition of a hydroxyl to the
sulfur atom and subsequent fragmentation of the unstable adduct [340]:

$$(CH_3)_2S + HO^. \longrightarrow (CH_3)_2SOH \longrightarrow \dot{C}H_3 + CH_3SOH \tag{5.102}$$

$$CH_3SOH + O_2 \longrightarrow CH_3SO_2OH \tag{5.103}$$

Methane sulfonic acid is a constant component of marine aerosols, and its
concentration near the equator in the central part of the Pacific Ocean generally
ranges from 9 to 75 ngm^{-3} [341,342]. It is a strong acid and can displace HCl
from NaCl forming a part of marine aerosol. It is highly soluble in water and is
oxidated in the liquid-drop phase with the formation of sulfuric acid which is
removed from the atmosphere with precipitation.

Atkinson R. et al. consider the reaction with a hydroxyl to be the main sink for
dimethyl sulfide during the daytime. According to their opinion, in the night the
main sink is the reaction with the NO$_3$ radical, although its rate constant is about
20 times lower than that for the reaction with a hydroxyl [343a].

5.4   Photochemistry of the Polluted Urban Atmosphere

A peculiar type of pollution of the urban atmosphere detected first in 1944 in Los
Angeles has been called "photochemical smog". In contrast to the well-known
"London smog", a dense fog with the admixture of particles of soot and sulphur
oxides, the photochemical smog appears as a result of the action of sunlight, most
often under the conditions of stable stratification of the atmosphere at low air
humidity. An indication of the formation of smog is the appearance of a bluish
haze and, as a result, a decrease in visibility. A marked irritation of the mucous

membranes of the respiratory tract and eye is observed. The maintenance of the smog situation for a long period leads to increasing morbidity and mortality of the population. Smog has a particularly strong effect on children and elderly people. It also has a detrimental effect on vegetation causing withering and decay of foliage. The more remote consequences are increasing metal corrosion and destruction of rubber and other materials.

The main compounds responsible for these properties of smog are ozone and peroxyacetylnitrate. It is these compounds that impart the oxidative character to the photochemical smog. In 1952, Haagen-Smith [343b] suggested that they are formed as a result of the action of sunlight from components of automobile exhaust gases. This suggestion was latter experimentally confirmed.

The dynamics of ozone accumulation in the atmosphere can be explained if the processes of transformation of nitrogen oxide under various conditions are considered. In the troposphere, the formation and dissociation of the ozone molecules occurs as a result of the following cyclic reactions:

$$NO_2 \xrightarrow{h\nu} NO + O(^3P) \quad (J = 7.2 \times 10^{-3}\,s^{-1}) \tag{5.30}$$

$$O(^3P) + O_2 + M \longrightarrow O_3 + M \tag{5.8}$$

$$O_3 + NO \longrightarrow NO_2 + O_2 \quad (k = 1.8 \times 10^{-14}\,cm^{-3}\ molecule^{-1}\,s^{-1}) \tag{5.104}$$

The decomposition of ozone in the reaction with $O(^3P)$ and $HO^{\cdot}$ in the troposphere proceeds slowly because of the low concentration of these particles. The rates of Eqs (5.30) and (5.104) are expressed by the equations

$$v_{(5.30)} = J_{(5.30)}\,[NO_2]$$

$$v_{(5.104)} = k_{(5.104)}[NO]\cdot[O_3]$$

Under stationary conditions we have $v_{(5.30)} = v_{(5.104)}$, and hence the ozone concentration is determined by the equation:

$$[O_3] = \frac{[NO_2]}{[NO]} \cdot \frac{J_{(5.30)}}{k_{(5.104)}} \tag{5.105}$$

Substitution into this equation of the value of $J_{(5.30)}$ ($7.2 \times 10^{-3}\,s^{-1}$), and the ratio of concentrations of NO_2 and NO typical of the urban air equal to 0.3 gives an ozone concentration of $1.2 \times 10^{11}\,cm^{-3}$ or about 5 ppb. This value is much lower than that observed even in the atmosphere of background areas.

Equation (5.105) shows that ozone concentration will increase with the rate of conversion of NO into NO_2. This acceleration is observed in the urban atmosphere as a result of the participation of hydrocarbons, carbonyl compounds and carbon oxide in the reactions. As shown in the previous sections of this chapter, the reaction of organic compounds with the hydroxyl radicals leads to consecutive reactions which can be described in the general form by Eqs. (5.36, 5.37, 5.38).

In the case of alkenes and alkanes the addition of hydroxyl or abstraction of hydrogen yield free radicals which subsequently react according to Eqs. (5.37) and (5.38). The photolysis of carbonyls and oxidation of CO lead to the peroxide radicals which also react rapidly with NO (Eqs. (5.27) and (5.77)).

The increase in the relative content of aromatic hydrocarbons should evidently lead to a certain decrease in the concentration of ozone and in its accumulation rate as a result of low yield of the peroxide radicals and the removal of a part of nitrogen oxides in the form of nitrophenols. The formation of alkylnitrates and nitrites, peroxyacyl nitrates and inorganic nitrogen compounds, such as water-soluble N_2O_5 and HNO_3, has a similar effect.

Hence, the accumulation of ozone depends on the ratio of initial concentrations of organic compounds, precursors of the peroxide radicals and nitrogen oxides. When this ratio is low, the rate of conversion of NO into NO_2 is also low and nitrogen oxide is involved in the process of ozone decomposition (5.104). At a very high ratio, ozone will not be accumulated either, first, because nitrogen dioxide is bonded by organic radicals and second, because of the reaction of the resulting O_3 with hydrocarbons.

The prediction of the behavior of complex photochemical systems under various conditions is of paramount practical importance for predicting smog situations and developing the strategy of struggle against photochemical pollution of the urban atmosphere. In recent years, considerable advances have been made by using the computer simulation of these systems with the aid of powerful computers. The development of a detailed model for ozone formation in the photochemical oxidation of even one individual compound requires more than one hundred elementary reactions to be taken into account. It is clear that even with the modern techniques of computer engineering this approach is inapplicable if it is necessary to represent the processes occurring in the urban atmosphere when many dozens of compounds and varying meteorological conditions and phenomena of transport and scattering are taken into account. Hence, we need to decrease drastically the volume of information entered into the computer concerning the chemistry of the process by using a reasonable simplification of the system under analysis. Two approaches are most often used for the solution of this problem [344].

The first of them is the replacement of an entire class of organic compounds by an individual "typical" representative with the inclusion into the model of a detailed mechanism for its oxidation. The second approach implies the application of a generalized oxidation mechanism to an individual class of organic compounds. This generalization is carried out on the basis of structural concepts and experiments in smog chambers with the representatives of the class under consideration. Each of these approaches has its own drawbacks. For example, the former approach requires a search for the criterion according to which a compound may be considered to be a typical representative of its class. It is often difficult to find this criterion and hence two or more compounds are included in the model. The same procedure is used in the latter approach when it is difficult to find a single mechanism for the oxidation of a great number of isomers and homologues forming this class of organic compounds.

On the whole, these simplifications proved to be very useful and in

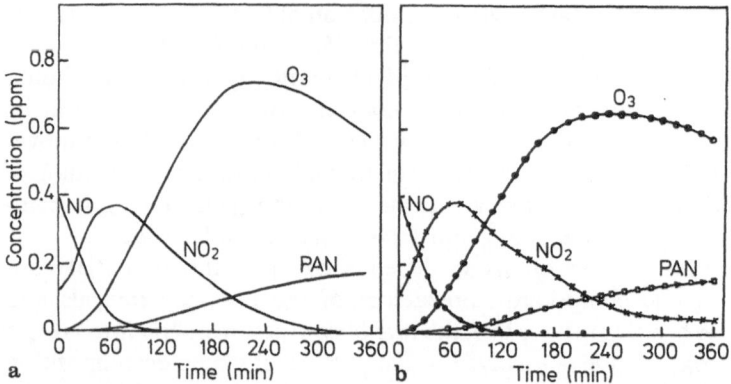

Fig. 5.4a, b. Calculated (**a**) and experimental (**b**) concentration profiles of some components in the oxidation in a smog chamber of a mixture of seven hydrocarbons [344]

experiments in smog chambers gave good agreements between the calculated and experimentally found concentration profiles of the initial compounds and the reaction products. An example of the application of computer simulation to the gas-phase oxidation of a mixture of seven organic compounds in the presence of nitrogen oxide is shown in Fig. 5.4.

The transition from mathematical modeling of photochemical reactions in smog chambers to that of processes in the open atmosphere requires information about the strength of the sources and the composition of organic components present in the atmosphere. This information should not be less reliable than the kinetic data available. This in turn requires the development of reliable and high-speed methods of controlling emission and atmospheric content of organic compounds.

Inferior visibility during smog is due to the formation of aerosol particles. The study of the mechanism for conversion of gaseous organic compounds into aerosols is important not only for solving the problem of smog formation. There is every reason to suppose that the formation of aerosols and their subsequent removal in the processes of impaction, dry and wet (with atmospheric rainfall) precipitation is one of the main paths of scavenging of the atmosphere. The level of current concepts of the mechanism for all these processes does not yet correspond to their importance.

Atmospheric organic aerosols can form according to a homogeneous or a heterogeneous mechanism. The former mechanism involves gas-phase oxidation of organic compounds, whereas the latter implies sorption, catalytic oxidation and, possibly, polymerization on the surface of already existing solid and liquid aerosols. Most researchers are inclined to think that homogeneous processes are the main processes for organic precursors. The relatively high vapor pressure of the main mass of organic components of the atmosphere and low solubility in the aqueous phase are presented as arguments in favor of this concept.

It is possible to formulate some requirements which should be met by organic compounds in order that their conversion into aerosol particles might take place. First, the molecules of these compounds should be readily oxidized and form products with a tendency towards condensation, i.e. the vapor pressure of oxidized products should be very low in order that their concentration in the air might exceed the pressure of the saturated vapor. It is clear that unsaturated and aromatic hydrocarbons meet the requirement of the facility of oxidation to a greater extent. The condition concerning the pressure of saturated vapor suggests that organic compounds with the number of carbon atoms from 5 to 10 should be of great interest. This range is due to the fact that upon oxidation the organic compounds with the number of carbon atoms in the molecule less than five yield mainly products with a high pressure of saturated vapor. On the other hand, the organic component of the atmosphere contains such a small fraction of compounds with more than ten carbon atoms and heavier than terpenes the molecular weight of which is 136, that they cannot be responsible for the formation of large amounts of aerosols even in very contaminated urban air.

A series of model experiments have been carried out confirming the fact that some hydrocarbons in the above range can readily be converted into aerosols [334, 345, 346a]. Fox D.L. et al. [346b] studied the rate of aerosol formation in photochemical reactions and in the reaction of aromatic hydrocarbons and cycloalkenes with ozone in the dark. A part of their results is presented in Table 5.2 and Fig. 5.5. Experiments were carried out in a smog chamber (with a volume of $200 \, m^3$) at night and in daylight. Aromatic hydrocarbons in the gas phase were found to be stable with respect to ozone, whereas cycloalkenes

Table 5.2 The rate of accumulation of aerosol particles in photochemical and dark smog-chamber reactions of some hydrocarbons

Hydrocarbon	Initial concentration, ppm	Maximal rate of conversion of hydrocarbon, $ppm \, h^{-1}$	Accumulation rate of aerosol particles, $\mu m^3 \, cm^{-3} \, h^{-1}$	
			$< 0.5 \, \mu m$	$> 0.5 \, \mu m$
		NO_x + hydrocarbon in day		
Toluene	2.70	0.17	110	8.0
m-Xylene	3.82	0.30	70	4.0
Cyclohexene	3.0	1.00	30	960
Cyclopentene	3.0	0.64	20	280
		Ozone + hydrocarbon at night		
Toluene	2.46	0.0[a]	2.0	0.7
m-Xylene	2.09	0.0[a]	3.0	0.9
Cyclohexene	2.02	0.91	150	270
Cyclopentene	1.59	0.99	22	73
Background	—	—	2.0	0.1

[a]Calculated values not significantly different from zero

actively reacted with it forming aerosol particles. In the day-time, toluene and xylene also actively participated in this process. Figure 5.6 shows that the concentration profiles of aerosols and ozone in the smog chamber are similar. An interesting fact which is difficult to interpret at present is the difference in fraction composition between aerosols formed by aromatic hydrocarbons and those formed by cycloalkenes.

It is not clear how the nucleation of oxidation products of hydrocarbons takes place. The particles already present in the atmosphere or stable complexes, formed by several tens of water molecules and called clusters, probably play an important role in this process. These clusters may be the condensation centres of vapors of low volatile aerosol-forming compounds. Clusters can also be formed by compounds other than water molecules. Condensation clusters consisting of carbon and silicates are formed in flue and exhaust gases. Finally, the molecules of aerosol-forming compounds themselves can form clusters on which nucleation of water and other components takes place. The relative contribution of processes of the formation of new particles and the capture of organic compounds by already existing particles depends on the rate of generation of condensed molecules in photochemical reactions, on their thermodynamic properties, and on the quantity of aerosols present in the system.

A paper by McMurry P.H. and Friedlander S.K. [347a] is concerned with the investigation of the role of aerosols already present in the atmosphere in the formation of new particles. These authors have shown (Fig. 5.6) that the higher the concentration of aerosols in the smog chamber (even before irradiation with light), the greater is the rate at which the products of hydrocarbon oxidation are involved in the process of heterogeneous nucleation. In our opinion, homogeneous processes play a limited role in the troposphere and are even more negligible in urban air because they require very high supersaturation. Vapor

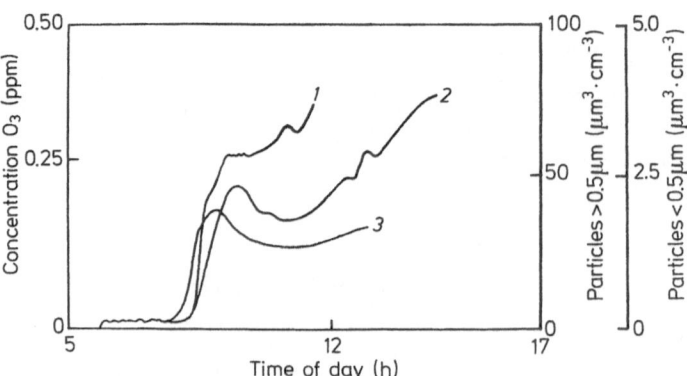

Fig. 5.5. Dynamics of the formation of aerosol particles and ozone upon the irradiation of a mixture of m-xylene, NO and NO_2 with sunlight. 1- Particles with d $<$ 0.1 μm, 2- particles with d $>$ 1 μm, 3- ozone

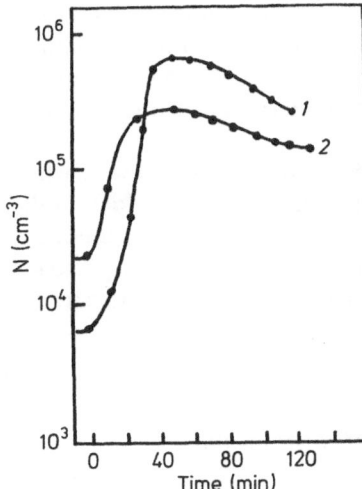

Fig. 5.6. Number of aerosol particles larger than 0.01 μm as a function of irradiation time of an SO_2–NO_2–propylene mixture at initial particle concentration of $6.7 \times 10^3 cm^{-3}$ (1) and $2.4 \times 10^4 cm^{-3}$ (2) [311]

pressure should exceed by at least twice the pressure of saturated vapor over a planar surface. In the opposite case the particles will be unstable. These conditions of supersaturation can probably occur only in the plumes of automobile exhaust gases, waste gases from industrial sources and, possibly, in the upper layers of the atmosphere which are characterized by low temperature and low aerosol content.

Among all the organic components, the rates of annual emission of which are many millions of tons, terpene hydrocarbons evidently display the greatest trend towards the formation of aerosols. This fact is confirmed by experiments in smog chambers (see Sect. 5.2). As early as at the beginning of the 1960s F. Went [242c] suggested that the bluish haze observed in summer time above the coniferous forests of the Rocky Mountains in the West of the USA is an aerosol formed as a result of photochemical oxidation of terpenes. More recently, the composition of atmospheric particles collected in these areas has been analyzed. These analyses showed that the main material of the aerosol is ammonium sulfate. However, as already mentioned, modern methods of determination of the chemical composition of atmospheric aerosols do not give an adequate idea about the amount of organic compounds present in these aerosols. Hence, the problem of conversion of terpenes into aerosols under natural conditions still remains unsolved.

The formation of aerosol particles in urban air is often believed to be associated with the participation of sulfur dioxide emitted in the burning of coal and oil products. Experiments in smog chambers show that combined oxidation of SO_2 and hydrocarbons leads to the formation of sulfinic acids

$$SO_2(^3B) + RH \longrightarrow RSO_2H \qquad (5.106)$$

The measurment of the quantum yield of this reaction gave very low values, which casts doubt on the importance of sulfinic acids in the formation of smog in the open atmosphere. It seems to us that the significance of sulfur oxide consists mainly in the fact that inorganic clusters and finest aerosols comprised of sulfuric acid drops and ammonium sulfate are formed during its photochemical oxidation. These clusters and aerosols serve as accelerators of nucleation of organic aerosol-forming compounds.

Several authors have reported that the rate of SO_2 conversion increases when small amounts of hydrocarbons and other organic compounds are added to the SO_2-pure air system. It might be assumed that the role of these compounds is reduced to the generation of a large number of the hydroxyl and peroxide radicals. Heterogeneous reactions with the participation of inorganic components dissolved in liquid aerosols evidently play an important role.

5.5   Chemistry of Halogenated Organic Compounds

Among numerous problems of atmospheric chemistry there has probably been no problem that has aroused such lively discussion as that of the effect of halogen-containing compounds on the ozone layer located in the stratosphere. This discussion has progressed far beyond research laboratories, profoundly alarmed the general public, has had repercussions in the most competent international organizations and has been considered in parliamentary and governmental circles of many countries. A Coordination Committee on the ozone layer (CCOL) was formed in the 1970s and is functioning at present in the framework of the UN environmental program (UNEP). The World meteorological organization has formed the International Committee on atmospheric ozone (ICAO). The investigations are sponsored by both international and governmental organizations and large private corporations.

This interest in the ozone problem is understandable: this allotropic oxygen form contained in the atmosphere in negligible amounts (the thickness of the ozone layer reduced to normal conditions is 2.5 to 3 mm in the equatorial regions and up to 4 mm at high latitudes) protects the biosphere from the destructive effect of ultraviolet radiation fom the Sun. Furthermore, the inversion layer of relatively warm air (see Fig. 1.1) formed as a result of exothermic decomposition of ozone protects the lower layers and the Earth's surface from cooling.

As already mentioned, the photochemical theory of ozone formation in the stratosphere was formulated by Chapman. The main reactions of this mechanism, (5.7b) and (5.9), make it possible to write the following equation for the change of equilibrium ozone concentration with altitude:

$$\frac{d[O_3]}{dt} = 2J_{(5.7b)}[O_2] - 2k_{(5.9)}[O][O_3] \qquad (5.107)$$

However, measurements carried out in the 1960s indicated differences between the observed and calculated concentrations, the former being lower than those predicted by the Chapman mechanism. This fact led to the elucidation of the nature of processes responsible for ozone depletion different from Eq. (5.9). Crutzen [347b] and, almost simultaneously, Johnston [347c] suggested that nitrogen oxides take part in the destruction of ozone and the formation of its stratospheric cycle. Nitrogen oxide (N_2O) (I) is a source of NO

$$N_2O \xrightarrow{h\nu} N_2 + O(^1D) \quad (< 230\,\text{nm}) \tag{5.12}$$

$$N_2O + O(^1D) \longrightarrow 2NO \tag{5.29a}$$

The catalytic cycle of ozone destruction is described by the equations

$$\frac{\begin{aligned}NO + O_3 &\longrightarrow NO_2 + O_2\\ NO_2 + O &\longrightarrow NO + O_2 (k = 3.3 \times 10^{-12}\,\text{cm}^3\ \text{molecule}^{-1}\,\text{s}^{-1})\end{aligned}}{O + O_3 \longrightarrow 2O_2} \tag{5.108}$$

Hence, a new term should be introduced into Eq. (5.107). It takes into account Eq. (5.108) which is the limiting reaction of the last cycle

$$\frac{d[O_3]}{dt} = 2J_{(5.7b)}[O_2] - 2k_{(5.9)}[O][O_3] - k_{(5.108)}[O_2][NO] \tag{5.109}$$

The contributions of the second and third terms in Eq. (5.109) may be compared if the values of concentrations and rate constants $k_{(5.9)}$ and $k_{(5.108)}$ are taken at a temperature prevailing at an altitude of 20 km (217 K). This substitution shows that dissociation of ozone proceeds more than seven times faster in the reaction with nitrogen oxide than without this reaction. In order to describe the ozone cycle more precisely, it is necessary to take into account the sinks of NO and NO_2 from the stratosphere and the role of the hydroxyl radical (Eq. (5.25)). However, the contribution of this process is probably not very substantial because the concentration of nitrogen oxide in the intermediate stratospheric layers is at least 500 times higher than that of HO$^{\cdot}$ radicals.

In addition to the photolysis of nitrogen oxide (I) the emission rate of which is profoundly affected by the intensity of use of nitrogenous fertilizers in agriculture, another source of NO in the stratosphere is provided by exhaust gases from supersonic planes and in recent years also by the American Shuttle program. Many researchers believe that when the intensity of flights in the stratosphere increases, the rate of ozone dissociation will increase markedly and this will have an unfavorable effect on the vegetation and the animal world of the planet.

In 1974 Molina M.J. and Rowland F.S. [148] indicated another danger for the ozone layer. They advanced a hypothesis that the ozone layer is destroyed as a result of the action of CFC's. The main theseis of Molina-Rowland's hypothesis may be formulated as follows:

(1) the emission of CF_2Cl_2 and $CFCl_3$ into the atmosphere is approximately equivalent to their world production;
(2) these compounds, extremely inert under the tropospheric conditions, slowly diffuse into the stratosphere;

(3) the photolytic decomposition of CFC's in the stratosphere leads to the emission of atomic chlorine involved in the catalytic cycle of ozone dissociation.

CFC's virtually do not participate in the reactions with the free radicals and are very slowly decomposed by light in the near ultraviolet region because of the high dissociation energy of C—Cl bond which is estimated to be $318\,kJ\,mole^{-1}$ for $CFCl_3$ and $337.7\,kJ\,mole^{-1}$ for CF_2Cl_2. Hence, the average time of their existence in the troposphere is many decades.

The study of the photolysis of $CFCl_3$ by ultraviolet light at the wavelength range 213.9 to 147 nm has shown that the primary process is the abstraction of one or two chlorine atoms

$$CFCl_3 \xrightarrow{h\nu} \dot{C}FCl_2 + Cl^\bullet \qquad (5.110a)$$

$$CFCl_3 \xrightarrow{h\nu} \ddot{C}FCl + 2Cl^\bullet \qquad (5.110b)$$

The total quantum yield of both reactions over the entire wavelength range indicated above is close to unity. However, their contribution differs greatly: upon irradiation with light at 213.9 nm, φ (5.110a) is 0.98, whereas in the vacuum, ultraviolet range abstraction of two chlorine atoms predominates ($\varphi_{(5.110b)}$ = 0.87 at λ = 147 nm).

The photolytic decomposition of CF_2Cl_2 takes place in a similar manner

$$CF_2Cl_2 \xrightarrow{h\nu} CF_2Cl + \dot{C}l \qquad (5.111a)$$

$$CF_2Cl_2 \xrightarrow{h\nu} \ddot{C}F_2 + 2\dot{C}l \qquad (5.111b)$$

The chlorine atoms formed in photolysis are involved in the cyclic process of ozone destruction:

$$Cl^\bullet + O_3 \longrightarrow ClO + O_2 \qquad (5.112)$$

$$\frac{ClO + O^\bullet \longrightarrow Cl^\bullet + O_2}{O_3 + O \longrightarrow 2O_2} \qquad (5.113)$$

Reactions (5.112) and (5.113) proceed very rapidly: their rate constants are $2.7 \times 10^{-11} \exp(-257/T)$ and $7.5 \times 10^{-11} \exp(-120/T)$, respectively, in the temperature range 220–300 K. They proceed much more rapidly than reactions (5.104) and (5.108) of the catalytic cycle of ozone decomposition by nitrogen oxide and hence the efficiency of the Cl–ClO cycle is higher.

Chain termination occurs in the reaction of catalytically active particles with methane, nitrogen dioxide and the peroxide radical

$$Cl^\bullet + CH_4 \longrightarrow HCl + CH_3^\bullet \qquad (5.114)$$

$$Cl^\bullet + HO_2^\bullet \longrightarrow HCl + O_2 \qquad (5.115)$$

$$ClO + NO_2 \longrightarrow ClONO_2 \qquad (5.116)$$

Chlorine passes again into the active form by the reaction

$$HCl + HO^{\cdot} \longrightarrow Cl^{\cdot} + H_2O \tag{5.117}$$

$$ClONO_2 \xrightarrow{h\nu} ClO + NO_2 \tag{5.118}$$

The following processes are also of considerable importance:

$$ClO + NO \longrightarrow Cl^{\cdot} + NO_2 \tag{5.119}$$

$$HO^{\cdot} + HO_2^{\cdot} \xrightarrow{h\nu} H_2O + O_2 \tag{5.120}$$

One of them relates to each other the cycles of chlorine and nitrogen oxide and the other controls the content of the $HO^{\cdot}$ radicals in the stratosphere.

A complete description of the catalytic cycle of ozone destruction due to the action of chlorine atoms generated by CFC's should include about 80 different reactions. Quantitative conclusions from Molina-Rowland's hypothesis can be obtained by computer simulation. The precision of these calculations is strongly dependent on the completeness and reliability of the initial data. This refers, first, to the information on the strength and spatial distribution of sources of CFC's, the rate of diffusion into the stratosphere and the kinetics of the key reactions. According to the investigations carried out by the National Academy of Sciences of the USA, the precision of measurements of rate constants for reactions (5.117) and (5.120) has the greatest influence on the results of calculations. The overall uncertainties of measurements of only nine key reactions (5.112)–(5.120) are responsible for more than six-fold variations in the calculated rates of a decrease in the amount of the stratospheric ozone.

The average time of residence of CFC's in the atmosphere is a critically important parameter. The existence of even a relatively small tropospheric sink can profoundly affect the results of calculations. If the rate of this sink is taken to be only 1% per year, this will lower the calculated value of the decrease in ozone content in the stratosphere approximately twice. At present, the lifetimes of $CFCl_3$ and CF_2Cl_2 in the stratosphere are taken to be 54 and 80 years, respectively. The introduction into some models of the τ value equal to 30 years would lead to a three-fold decrease in the calculated reduction of the O_3 content.

CFC's are the most important but not the only source of the chlorine atoms destroying ozone. The evaluation reported in Sect. 4.2.3 shows that about 0.57 Tg of chlorine contained in $CFCl_3$ and CF_2Cl_2 is emitted into the stratosphere annually, which represents approximately 82% of the total emission from anthropogenic sources. The same authors think that another 0.24 Tg of chlorine contained in CH_3Cl is emitted into the atmosphere from natural sources. Naturally, all models should take into account not only the CFC's emission but also the photochemistry of other chlorine-containing compounds.

Among these compounds, apart from methyl chloride, CCl_4 and methyl chloroform, CH_3CCl_3, attract great attention. There are reasons to suppose that the most part of CCl_4 emitted into the atmosphere sooner or later will enter the stratosphere. The absorption spectrum of the CCl_4 molecule becomes continuous in the near ultraviolet range at about 250 nm. As in the case of CFM's, the process takes mainly two directions.

$$CCl_4 \xrightarrow{h\nu} \dot{C}Cl_3 + Cl^{\cdot} \qquad (5.121a)$$

$$CCl_4 \xrightarrow{h\nu} \ddot{C}Cl_2 + 2Cl^{\cdot} \qquad (5.121b)$$

with the total quantum yield close to unity. The dissociation proceeding according to Eq. (5.121b) becomes important in vacuum ultraviolet ($\lambda < 200$ nm).

In photolysis, the incompletely halogenated hydrocarbons (CH_3Cl, $CHCl_3$, CH_3CCl_3, etc.) abstract a chlorine atom but not a hydrogen atom. For example, methyl chloride absorbing only the light at a wavelength shorter than 200 nm dissociates according to the scheme

$$CH_3Cl \xrightarrow{h\nu} CH_3^{\cdot} + Cl^{\cdot} \qquad (5.122)$$

However, in contrast to $CFCl_3$ and CF_2Cl_2, these compounds exhibit a considerable sink into the troposphere because they can interact with the hydroxyl radicals.

The rate constants for the reaction of halocarbons with the $HO^{\cdot}$ radicals are relatively low and range from 1 to 5×10^{-14} cm^3 molecule^{-1} s^{-1}. Among the oxidation products of methyl chloride, HCl, carbon monoxide and unstable formyl chloride have been detected [348]. They are formed by the following reactions:

$$CH_3Cl + HO^{\cdot} \longrightarrow \dot{C}H_2Cl + H_2O \qquad (5.123)$$

$$\dot{C}H_2Cl + O_2 \longrightarrow {}^{\cdot}OOCH_2Cl \qquad (5.124)$$

$${}^{\cdot}OOCH_2Cl + NO \longrightarrow {}^{\cdot}OCH_2Cl + NO_2 \qquad (5.125)$$

$${}^{\cdot}OCH_2Cl + O_2 \longrightarrow HC{\overset{\displaystyle O}{\underset{\displaystyle Cl}{\big<}}} + HO_2^{\cdot} \qquad (5.126)$$

The photolysis of formyl chloride yields HCl and CO

$$HCOCl \xrightarrow{h\nu} CO + HCl \qquad (5.127)$$

The photochemical oxidation of eight chloro-ethanes initiated by chlorine atoms has been investigated in ref. [348]. In addition to HCl, CO and HCOCl, the reaction mixture was found to contain phosgene. Its formation is also observed in the oxidation of chloroform. According to Singh's report [86], phosgene has been found in the air of cities on the western shore of the USA at concentrations of 30–60 ppt. Its formation in the atmosphere from methylchloroform may be represented by the following sequence of reactions

$$CCl_3CH_3 \xrightarrow{HO^{\cdot}} CCl_3\dot{C}H_2 \xrightarrow{O_2} CCl_3CH_2OO^{\cdot} \xrightarrow{NO} \qquad (5.128)$$

$$\longrightarrow CCl_3CH_2O^{\cdot} \longrightarrow \dot{C}Cl_3 + CH_2O$$

$$\dot{C}Cl_3 \xrightarrow{O_2} {}^{\cdot}OOCCl_3 \xrightarrow{NO} {}^{\cdot}OCCl_3 \xrightarrow{O_2} Cl\!-\!\underset{\underset{O}{\|}}{C}\!-\!Cl + ClO_2^{\cdot} \qquad (5.129)$$

The phosgene molecule absorbs sunlight in the near ultraviolet region of the spectrum (238–305 nm) dissociating into CO and the chlorine atoms which can recombine forming the Cl_2 molecule.

It has repeatedly been emphasized in this book that in many atmospheric processes a number of organic compounds present in the troposphere and stratosphere in quite negligible amounts are of tremendous importance. One of the most outstanding examples of the effect of microcomponents on the global processes of formation of the radiation balance and the climate of the planet is provided by the stratospheric chemistry of organobromine compounds. Many researchers have insistently directed the attention of specialists in the chemistry of the atmosphere and geophysics to the peculiar role played by the bromine atoms in the chemical processes occurring in the lower layers of the stratosphere [349a].

All the one-dimensional models serving for the study of the effect of CFC's on ozone indicate that this effect is most pronounced in the stratospheric layer above 25 km. In the lower stratosphere in which the rates of regeneration processes of nitrogen monoxide (5.108) and chlorine atoms (5.119) limiting the rate of the overall process of ozone destruction are low, the bromine atoms play the most important role.

Bromine is emitted into the stratosphere mainly in the form of trifluorobromomethane (Halon 1301) and difluorobromochloromethane (Halon 1211), the time of existence of which in the air layers near the ground level is estimated to be about 70 years. The stratospheric chemistry of bromine is evidently related to a lesser extent to the penetration of methyl bromide and tribromomethane as well as dibromomethane and bromochloromethane through the tropopause. The residence time of the two former compounds in the troposphere is estimated to be 1–2 years. Both natural and anthropogenic sources of the above bromine compounds have low strength and therefore their concentrations are very low: they are usually lower by one or two orders of magnitude than in the case of their chlorine-containing analogues (see chap. 2).

The bromine-containing Halons are readily decomposed upon irradiation at the near ultraviolet wavelength

$$CBrF_3 \xrightarrow{h\nu} CF_3^{\cdot} + Br^{\cdot} \qquad (5.130)$$

$$CBrF_2Cl \xrightarrow{h\nu} \dot{C}F_2Cl + Br^{\cdot} \qquad (5.131)$$

Wofsy S. C. et al. [349b] have postulated the participation of bromine atoms in two catalytical cycles of ozone destruction

$$Br^{\cdot} + O_3 \longrightarrow BrO + O_2 \tag{5.132}$$

$$\frac{BrO + O^{\cdot} \longrightarrow Br^{\cdot} + O_2}{O^{\cdot} + O_3 \longrightarrow 2O_2} \tag{5.133}$$

and

$$2(Br^{\cdot} + O_3 \longrightarrow BrO + O_2) \tag{5.132}$$

$$\frac{BrO + BrO \longrightarrow 2Br^{\cdot} + O_2}{2O_3 \longrightarrow 3O_2} \tag{5.134}$$

The bromine atoms pass into the inactive form, HBr, as a result of the processes:

$$Br^{\cdot} + HO_2^{\cdot} \longrightarrow HBr + O_2 \tag{5.135}$$

$$Br^{\cdot} + H_2O_2 \longrightarrow HBr + HO_2^{\cdot} \tag{5.136}$$

$$Br^{\cdot} + H_2CO \longrightarrow HBr + H\dot{C}O \tag{5.137}$$

Reactions with methane and molecular hydrogen playing an important role in the processes of deactivation of atomic chlorine are not of great importance in the case of the stratospheric chemistry of bromine because they proceed at a low rate.

The regeneration of atomic bromine occurs in the interaction of HBr with the hydroxyl radical and atomic oxygen and in photolytic decomposition:

$$HBr + HO^{\cdot} \longrightarrow Br^{\cdot} + H_2O \tag{5.138}$$

$$HBr + O^{\cdot} \longrightarrow Br^{\cdot} + HO^{\cdot} \tag{5.139}$$

$$HBr \xrightarrow{h\nu} Br^{\cdot} + H^{\cdot} \tag{5.140}$$

At temperatures above 250 K, the rate constant for reaction (5.138) is $8.5 \times 10^{-12}\,cm^3\,molecule^{-1}\,s^{-1}$ and that for reaction (5.139) is $7.6 \times 10^{-12} \exp(1571/T)\,cm^3\,molecule^{-1}\,s^{-1}$. The photodissociation coefficient $J_{(5.140)}$ at an altitude of 30–40 km and a latitude 30 °N is $(4.9–5.8) \times 10^{-6}\,s^{-1}$.

It should be noted that the [Br]/[HBr] ratio in the stratosphere is higher than the [Cl]/[HCl] ratio. This difference is due to the relatively low rate of HBr formation as compared to that of HCl and to the relatively high rate of processes (5.139)–(5.140). The BrO species generated in Eq. (5.132) has low activity. In this case the regeneration of atomic bromine occurs as a result of reactions (5.133) and (5.134) and the following processes:

$$BrO \xrightarrow{h\nu} Br^{\cdot} + O^{\cdot} \tag{5.141}$$

$$BrO + NO \longrightarrow Br^{\cdot} + NO_2 \tag{5.142}$$

$$BrO + HO^{\cdot} \longrightarrow Br^{\cdot} + HO_2^{\cdot} \tag{5.143}$$

$$BrO + O_3 \longrightarrow Br^{\cdot} + 2O_2 \tag{5.144}$$

$$BrO + ClO \longrightarrow Br^{\cdot} + Cl^{\cdot} + O_2 \tag{5.145a}$$

$$\longrightarrow Br^{\cdot} + ClO_2 \tag{5.145b}$$

The photolytic decomposition of BrO proceeds much faster than that of ClO According to Yung Y. L. et al. [349], the photodissociation coefficients for ClO and BrO differ by two or three orders of magnitude.

Taking into account this reaction and also some other reactions, Yung Y. L. et al. [349] have added four other cycles to the two cycles of ozone dissociation with the participation of bromide atoms postulated by Wofsy S. C. et al. [349b].

The first of them

$$Br^{\cdot} + O_3 \longrightarrow BrO + O_2 \tag{5.132}$$

$$Cl^{\cdot} + O_3 \longrightarrow ClO + O_2 \tag{5.112}$$

$$\frac{BrO + ClO \longrightarrow Br^{\cdot} + Cl^{\cdot} + O_2}{2O_3 \longrightarrow 3O_2} \tag{5.145a}$$

relates to each other the catalytic cycles of bromine and chlorine compounds. This sequence of reactions leads to an important conclusion that in the lower stratosphere the efficiency of the catalytic cycle of ozone destruction by chlorine atoms increases in the presence of bromine atoms. Hence, even a slight increase in the emission of bromine compounds into the stratosphere is of great potential danger to the ozone layer.

The second cycle relates the chemistry of bromine compounds to that of nitrogen oxides

$$Br^{\cdot} + O_3 \longrightarrow BrO + O_2 \tag{5.132}$$

$$BrO + NO_2 + M \longrightarrow BrONO_2 + M \tag{5.146}$$

$$BrONO_2 \xrightarrow{h\nu} Br^{\cdot} + NO_3 \tag{5.147}$$

$$NO_3 \xrightarrow{h\nu} NO + O_2 \tag{5.148}$$

$$\frac{NO + O_3 \longrightarrow NO_2 + O_2}{2O_3 \longrightarrow 3O_2} \tag{5.104}$$

Here the key reactions are those of the formation and photodissociation of $BrONO_2$. If alternative directions of the photolytic dissociation predominate in the stratosphere

$$BrONO_2 \xrightarrow{h\nu} BrO + NO_2 \tag{5.149a}$$

$$\xrightarrow{h\nu} BrONO + O(^3P) \tag{5.149b}$$

then this cycle is degenerate, i.e., it does not affect the overall process of ozone destruction.

The third cycle takes into account reactions with the hydroxyl and hydroperoxide radicals

$$Br^{\cdot} + O_3 \longrightarrow BrO + O_2 \tag{5.132}$$

$$BrO + HO_2^{\cdot} \longrightarrow HOBr + O_2 \tag{5.150}$$

$$HOBr \xrightarrow{h\nu} HO^{\cdot} + Br^{\cdot} \tag{5.151}$$

$$\frac{HO^{\cdot} + O_3 \longrightarrow HO_2^{\cdot} + O_2}{2O_3 \longrightarrow 3O_2} \tag{5.25}$$

The limiting stage of the cycle is reaction (5.150) for which according to the evaluation of Yung Y. L. et al., the rate constant is less than $4 \times 10^{-2} \, cm^3$ molecule^{-1} s^{-1} at 298 K.

The role of the fourth cycle

$$Br^{\cdot} + O_3 \longrightarrow BrO + O_2 \tag{5.132}$$

$$\frac{BrO + O_3 \longrightarrow Br^{\cdot} + 2O_2}{2O_3 \longrightarrow 3O_2} \tag{5.144}$$

in ozone dissociation is evidently minor becuse the rate of reaction (5.144) at a temperature typical of the lower stratosphere (about 220 K) does not exceed $6 \times 10^{-16} \, cm^3$ molecule^{-1} s^{-1}.

According to the authors of ref. [349], the total effect of bromine atoms on the chemical processes should be expressed not only in a decrease in ozone concentration in the lower layers of the stratosphere but also in that at the temperature of the Earth's surface.

5.6 Some Photostimulated Reactions on Aerosol Surfaces

Sections 5.2–5.5 have been concerned exclusively with homogeneous gas-phase processes of oxidation of organic atmospheric components. The data reported there on the rates of photo- and thermochemical processes and the composition of the products have been obtained as a result of laboratory investigations in smog chambers filled with artificial gas mixtures with a known composition thoroughly freed from impurities. However, the real atmosphere is a complex heterogeneous system containing a large amount of dispersed substances: solid and liquid aerosol particles.

The principal natural sources emitting the major part of aerosols are volcanoes, oceans, rock weathering and forest fires. These sources emit aerosols greatly differing in composition.

Volcanic ash is a silicate material. However, apart from SiO_2 it contains relatively large amounts of rare earth elements (Sr, Ta, Yb, Sc, etc.) as well as iron, chromium, copper and zinc. The composition of volcanic ash collected during eruptions of two volcanoes of Kamchatka and the Kurile Islands [4a] is given below in per cent:

Volcano, year	SiO_2	TiO_2	Al_2O_3	Fe_2O_3	FeO	MgO	C_2O
Tyatya, 1973	50.3 –	0.8 –	13.8 –	3.4 –	0.7 –	3.5 –	8.5 –
	– 53.8	– 1.4	– 19.0	– 11.6	– 8.0	– 5.5	– 9.8
Tolbachik, 1976:							
Southern crater	50.9	1.8	16.5	3.9	6.7	5.0	8.4
Northern crater	50.0	1.0	13.4	3.9	5.8	9.8	12.0

The main mass of eruption products of land volcanoes consists of solid particles with a wide range of sizes, formed as a result of magma dispersion. In the process of an explosive eruption, an eruptive cloud is formed which often attains a height of tens of kilometers. As a result of these eruptions, aerosols reach not only the upper troposphere but also the stratosphere. For example, the products of an eruption of the El Chichon volcano in March–April 1982 (zinc and copper oxides) have been detected at altitudes of 18–21 km [350].

The data obtained in the last 10–15 years by volcanologists in various countries show that the major part of the erupted material consists of ash particles including the finest particles ranging from 0.01 to 1 μm. The latter are stable with respect to gravitational sedimentation. The characteristic time of their sedimentation from the troposphere is many tens of days, and hence they are transported over long distances. Until recently, in the determination of the mass of the erupted material, only particles with the size of 1 μm and higher have been taken into account. According to the evaluation of Zemtsov A. N. [351], the "latent mass" of the finest long-lived ash particles is not less than 10–20% of the erupted mass. These particles are emitted into the atmosphere at a rate of 0.5–1 $km^3 y^{-1}$, and their total surface area exceeds $10^{16} m^2$ [351].

Marine aerosol is mainly formed when bubbles on wave crests are destroyed and contains NaCl as the main component. However, it also contains many compounds of other metals. According to some evaluations, oceans provide 5–20% of the total amount of copper, vanadium, and zinc compounds present in the atmosphere [352]. The emission of iron zinc and copper into the atmosphere by dispersion of sea water is estimated to be 2.6×10^6, 1.4×10^6 and $1.7 \times 10^5 ty^{-1}$, respectively [353a].

Dust stroms in deserts and wind erosion of soils are also considerable sources of aerosols. Great masses of dust are often transported from large deserts over tremendous distances: for example, the dust from Sahara is precipitated in the countries of Central Europe and on the coast of South America. According to Junge [353b], the Sahara dust has the following chemical composition: SiO_2, 37–55%, Al_2O_3 up to 20%, Fe_2O_3 1–16%, MnO 2–4%, $CaCO_3$ 6–22% and MgO 0.4–4%.

During weathering, oxides and sulfides of many metals forming a part of rocks and minerals enter the atmosphere. For example, titanium is present in the form of ilmenite (TiO_2 52.7%, FeO 47.3%), rutile and anatase (94–99% TiO_2). Iron oxides and sulfides, zinc oxide (sphalerite), lead oxide (galenite) and tin oxide (cassiterite) are widely spread.

Hence, volcanic eruptions, dust and sand storms, choppy sea and forest fires transform the Earth's troposphere into a colloid system with a very high free energy value.

It should be noted that oxides and sulfides of many metals exhibit the properties of semiconductors and probably play an important role in the sinks for organic components of the atmosphere according to the adsorption–sedimentation and photocatalytic mechanisms.

Photocatalytic transformations of organic compounds are being intensively investigated by specialists in the field of environmental chemistry. The greatest attention is being devoted to the study and practical use of photocatalytic dissociation of complex organic substances in aqueous solutions in connection with the problems of sewage treatment. Many authors have indicated that photocatalytic processes may also have a certain effect on the chemistry of the atmosphere, but investigations in this field are only just beginning. Hence, at present we may only speak of the potential significance of photocatalytic reactions in the formation of sinks for minor gaseous components of the atmosphere.

The interpretation of the mechanism for photocatalytic processes with the participation of metal oxides and sulfides may be found in the framework of the so-called zone theory of semiconductors. According to this theory, valence electrons of atoms in a crystal form a single system in which two electrons with the same set of quantum numbers and hence with the same energy cannot exist (Pauli's principle). Consequently, in the formation of a crystal, the energetic levels of electrons are split into N levels (Scheme 5.7). The normal electronic level has the lowest energy E_0. This splitting is observed not only for the ground state but also for the excited state of electrons. Hence, energetic zones divided into levels exist in a crystal. In the crystal of a semiconductor, all the levels of the ground state (valence zone) are unavailable, and within this zone no changes in the state of electrons are possible including the migrations due to the action of the electric field. For the appearance of conduction in a crystal, a part of electrons should pass into the zone of levels $E_1^* - E_1^1$ (conduction zone) in which they behave as free electrons in metals. This zone is separated from the valence zone by the region of forbidden states in which electrons cannot exist. The width of this forbidden zone ($E_1^* - E_0^1 = E_G$) differs for different semiconductors. For semiconductors, electric conduction is determined by the equation

$$\sigma = \sigma_0 \exp(-E_a \cdot k^{-1} \cdot T^{-1}) \qquad (5.145)$$

where k is Boltzmann's constant, E_a is the activation energy of electrons and σ_0 is the proportionality coefficient. This equation shows that under the influence of external energy sources, a part of electrons proportional to $\exp(-E_a k^{-1} T^{-1})$ passes into the conduction zone.

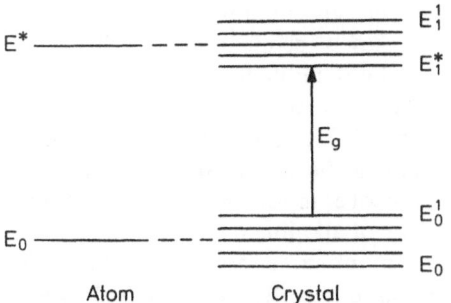

Fig. 5.7. Diagram of the ground and excited energetic levels in atoms and crystals of semiconductors

Hence, two electricity carriers appear in the semiconductor: the negatively charged electron (e^-) in the conduction zone and the positively charged vacancy, a hole (h^+), in the valence zone. In the pure crystal of a conductor, E_a cannot be smaller than E_g, and at a large width of the forbidden zone only a small part of electrons can cross it. However, if a crystal contains impurities the levels of valence electrons of which are not completely unavailable and lie within the forbidden zone $E_0^1 - E_1^*$, then the electrons of the valence zone can pass to these levels leaving the places in the zone $E_0 - E_0^1$ and ensuring the hole conduction of the semiconductor.

Impurities can also induce electron conduction if a transition from the levels of valence electrons of the atoms of impurities into the conduction zone of the semiconductor occurs. The significance of these impurities is very great: being present in negligible amounts they can change the electric conductivity of semiconductors millions of times.

The transitions of electrons into the conduction zone can be induced not only by thermal energy but also by the absorption of light quanta at a corresponding wavelength. In a pure crystal, the transition is induced by the light the quantum energy of which is not less than E_g. However, impurity photoconduction is ensured by the absorption of longer wavelength light as well. This point is very important for the photostimulated processes in the troposphere considered here.

Table 5.3. Width of the forbidden zone (E_g) and the threshold value of the wavelength of light exciting photoconduction in some semiconductors

Compound	E_g, eV	λ min, nm	Compound	E_g, eV	λ min, nm
Cr_2O_3	1.90	625.5	SiO_2	8.0	155.0
NiO	1.95	635.8	PbS	0.41	3023.9
V_2O_5	2.1	590.4	ZnS	3.6	344.4
Fe_2O_3	2.3	539.4	KI	6.0	206.6
TiO_2 (rutile)	3.1	399.9	KBr	7.4	167.5
ZnO	3.2	387.4	NaBr	7.5	165.3
SnO_2	3.5	354.2	KCl	8.4	147.6
γ-Al_2O_3	7.3	169.8	NaCl	8.5	145.8

As already mentioned, the atmospheric aerosol contains large amounts of inorganic compounds exhibiting the properties of semiconductors. Table 5.3 shows the values of E_g and wavelengths of light inducing the appearance of photoelectrons. It can be seen that among the most widespread aerosol components, only Fe_2O_3 absorbs light in the visible range. The absorption of TiO_2 and ZnO occurs at the boundary between the visible and the ultraviolet regions of the spectrum, whereas the main aerosol components (Al_2O_3 and SiO_2) absorb only in the far UV range. Halogenides of alkaline metals also absorb in this range. However, pure silicium and aluminium oxides and pure halogenides virtually do not exist in nature. It may be assumed that all these substances contain some amounts of impurities, and as a result photoelectrons and holes are generated in them upon irradiation with light in the near UV or even the visible range.

The non-equilibrium distribution of electrons, e^-, and holes, h^+, induces the redox reactions of molecules of compounds adsorbed on the surface of semiconductor particles. The driving force of chemical processes on photoexcited semiconductors is the electron transfer of the charge from the particle surface to the adsorbent or vice versa. The electron-acceptor molecules present on the surface react with photoelectrons, and as a result photochemosorption is observed, as, for example, in the case of oxygen on the surface of ZnO:

$$O_2 + e^- \longrightarrow O_{2(ads)}^- \tag{5.146}$$

or dissociation occurs characteristics of alkyl halogenides:

$$RHal + e^- \longrightarrow RHal^- \longrightarrow R^{\cdot} + Hal_{(ads)}^- \tag{5.147}$$

Electron-donor adsorbates interact with holes to give radicals:

$$RCOO^- + h^+ \longrightarrow R^{\cdot} + CO_2 \tag{5.148}$$

Photosorbed oxygen can also interact with a hole and pass into the atomic state:

$$O_{2(ads)}^- + h^+ \longrightarrow 2O_{(ads)} \tag{5.149}$$

Hence, the irradiation of semiconductor materials with light at a certain wavelength leads to the appearance of active centers, ions and radicals of various types on their surface.

The probability of the occurrence of some redox reactions on the surface of the semiconductor depends not only on the width of the forbidden zone and the energy of the incident light but also on the mutual arrangement of the valence zone and the conduction zone and on the redox potentials of the sorbates. Figure 5.8 is a diagram showing the zones of a semiconductor and the energetic levels of oxidation and reduction of the component. For reduction, the energetic level A^+/A should lie over the lower boundary of the conduction zone. Oxidation is observed if the B^-/B level is located over the upper boundary of the valence zone. For example, this arrangement of energetic zones with respect to redox levels of the process of water decomposition ($\Delta E^\circ = 1.23$ V) is characteristic of

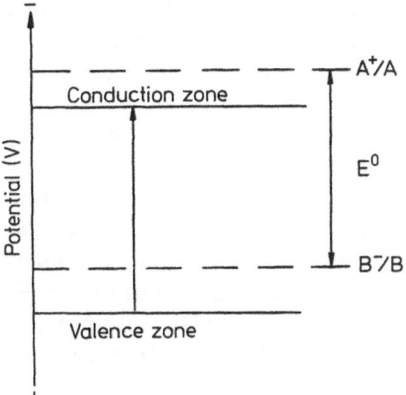

Fig. 5.8. Energetic levels of a semiconductor and redox levels of a reagent

TiO_2 (the potentials of the conduction zone and the valence zone with respect to normal hydrogen electrode in a solution with pH = 1 are − 0.1 and + 3.1 V, respectively). Titanium dioxide decomposes water to hydrogen and oxygen. Hence, the knowledge of the values of redox potentials of the reagents makes it possible to predict the trend of their transformations on the surface of irradiated semiconductors. Unfortunately, very little information is available in the literature about these potentials of organic compounds [354].

Up to the present, relatively many investigations have been carried out concerning photostimulated transformations of organic compounds on semiconductor materials. However, only a few of them have a direct relationship to atmospheric chemistry. Nevertheless, even these few papers suggest that the role of aerosol particles in the processes of photochemical transformations of organic components of the atmosphere is very important.

Very interesting results have been obtained by a team of Japanese authors in the study of the oxidation of isomeric C_4H_8 butenes in the presence of NO_2. It was found that the introduction of finely disperse zinc oxide into the photochemical reactor leads to a drastic change in the distribution of reaction products (aldehydes and ketones, epoxides, nitrates, HNO_3, CO and CO_2). Moreover, new compounds were recorded in the reactor. They contained a cyan group (HCN and CH_3CN) and are not formed in the absence of zinc oxide [355, 356]. The authors suggest that ZnO is manifested mainly in the reaction with products formed as a result of homogeneous processes. They have shown in special experiments that extensive photocatalytic oxidation of these products occurs to give CO_2.

As early as the end of the 1970s it was established that when organic compounds sorbed by various silicate materials were irradiated with light in the near UV region, they undergo extensive oxidation to give CO_2 [357–359]. Aliphatic and aromatic hydrocarbons as well as some of their oxygen- and halogen-containing derivatives were among the investigated compounds. Soon after this, the first attempt was made to evaluate the emission of CO_2 to the atmosphere as a result of photostimulated oxidation of organic compounds sorbed by the silicate materials of the Earth's crust (in these experiments, sand from the shore of California was used containing 3% Fe, 10%, Al, 5% Ca and 0.6 Ti) [360]. According to the opinion of the authors, this source may be responsible for the formation of 1.8×10^{15}g of CO_2 per year, which is about 30% of the anthropogenic emission due to the combustion of fossil fuel.

As can be seen from the E_g values given in Table 5.3, SiO_2 does not exhibit its own absorption in the near UV and visible spectral ranges. Hence, the photostimulated destruction of organic compounds observed by the authors of the above papers should be interpreted by the presence of impurities in silicates; iron and titanium compounds, etc. It has recently been shown that mixed oxides exhibit a higher photocatalytic activity than pure oxides [361, 362]. In particular, a considerable increase in TiO_2 activity in the reaction of photooxidation of unsaturated hydrocarbons was observed when it was introduced in the form of an admixture (4–10%) into SiO_2.

The information according to which metal oxides decompose even such stable compounds as CCl_4 and CFC's upon prolonged irradiation deserves particular attention [363–365]. In ref. [363] the photoinduced decomposition of CCl_4, $CFCl_3$ and CF_2Cl_2 on SiO_2 in the presence of ethane by irradiation at 366–405 nm wavelengths was investigated. The appearance of C_2H_5Cl in the gas phase showed that adsorbed molecules of methyl halogens undergo photodissociation.

Soviet authors [364] investigated by IR spectroscopy the interaction of the same methyl halogens with the surface of Al_2O_3 stimulated by UV radiation. They recorded changes in absorption spectra assigned to the appearance on the surface of Al_2O_3 of new compounds (CO_2 and phosgene) adsorbed on coordinatively unsaturated aluminium atoms. Furthermore, when a mixture of oxygen and CF_2Cl_2 was irradiated over Al_2O_3, the spectra of the gas phase exhibited absorption lines assigned to CF_3Cl and CCl_4. The authors of ref. [364] have suggested that the disproportionation of CF_2Cl_2 proceeds according to the following scheme:

$$3\,CF_2Cl_2 \xrightarrow[Al_2O_3]{h\nu} 2\,CF_3Cl + CCl_4 \qquad (5.150)$$

Filby W.G. et al. [365] investigated by X-ray photoelectron spectroscopy the behavior of CF_2Cl_2 and $CFCl_3$ on the surface of ZnO upon irradiation with light at 250–450 nm. The dissociation of these CFC's led to the chlorination of the ZnO surface. In order to describe these processes, a scheme was proposed involving the formation of photoelectrons on the surface of ZnO, the photoadsorption of oxygen and charge transfer from the surface to the CFC molecule:

$$ZnO \xrightarrow{h\nu} (ZnO^+, e^-) \qquad (5.151)$$

$$(ZnO^+, e^-) + O_2 \longrightarrow ZnO(O_2^-) \qquad (5.152)$$

$$(ZnO^+, e^-) + CFCl_3 \longrightarrow ZnO + CFCl_3^- \qquad (5.153)$$

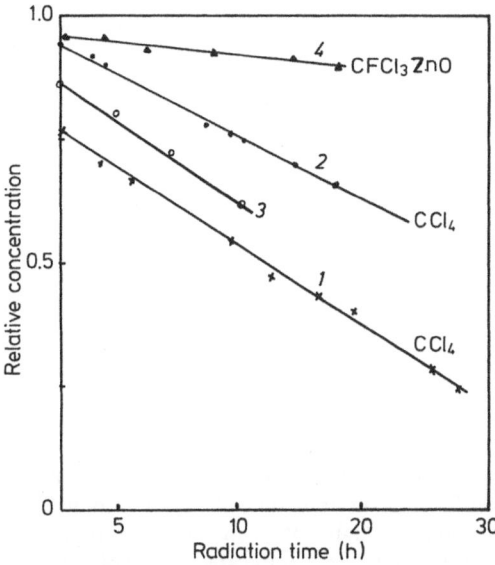

Fig. 5.9. Photocatalytic decomposition of CCl_4 on the surface of sand (*1*), volcanic ash (*2*), Fe_2O_3 (*3*), and decomposition of $CFCl_3$ on the surface of zinc oxide (*4*)

$$ZnO(O_2^-) + CFCl_3 \longrightarrow ZnO(O_2) + CFCl_3^- \tag{5.154}$$

$$CFCl_3^- \longrightarrow Cl^- + CFCl_2^{\cdot} \tag{5.155}$$

The author of the present book in collaboration with Klokova E.M. and Zgonnik P.V. investigated the photoinduced dissociation of CCl_4 and $CFCl_3$ on the surface of ZnO and some natural materials, sand from the Pacific coast of Mexico and volcanic ash particles collected during the eruption of the Tolbachik volcano (1976) on Kamchatka*. Natural materials consisting of particles smaller than 0.05 mm had previously been washed to remove chlorine ions and dried at 300–500 °C. Experiments were carried out in 500–550 ml glass reactors on the walls of which 3 to 8 g of the investigated material was fixed. A 250 W mercury lamp, placed in a quartz or pyrex tube and cooled with water, was introduced into the reactor. The initial concentration of methane halide was 1–15 ppm. The decrease in the amount of methane halide and the appearance of reaction products was checked by gas chromatography. The chlorine ion content in the catalyst was determined after the completion of the experiment by potentiometric titration with a chlorine-silver electrode. Preliminary results of investigations are shown in Figs. 5.9 and 5.10.

First, a considerable sorption (20–30%) of methane halide by the surface of the sand was observed. Volcanic ash sorbs CCl_4 to a lesser extent (8–10%). It is clear from Fig. 5.9 that during 15–20 h of irradiation at a wavelength of more than 300 nm, the CCl_4 concentration decreased by 25–30%. Phosgene and CO_2 were recorded among the oxidation products, and when irradiation was carried out with unfiltered light, Cl_2 also accumulated in the reactor.

---

*Isidorov VA, Klockova EM, Zgonnic PV (1990) Vestnik LGU. Ser 4, No. 1, p. 52

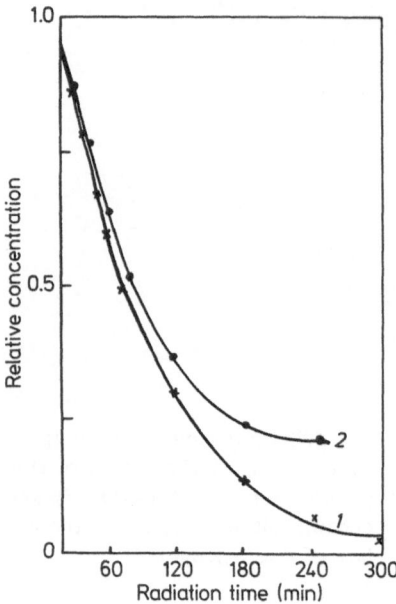

Fig. 5.10. Dependence of the relative content of CCl$_4$ on the time of irradiation with a complete light of a mercury lamp in the presence (*1*) and in the absence (*2*) of volcanic ash

Our experiments and the results of previous investigations [364, 365] convince us that a tropospheric sink for CCl$_4$ and CFC's exists: it is photo-stimulated decomposition on the surface of natural aerosols. The results of photolysis after irradiation with unfiltered light of the mercury lamp (Fig. 5.10) show that the processes are accelerated in the presence of ash particles. This effect suggests that atmospheric aerosols also play a certain role in the formation of a sink for CCl$_4$ and CFC's from the stratosphere. Finest particles of volcanic ash and sand can probably ensure the removal of a part of chlorine and prevent its inclusion into the catalytic cycle of ozone depletion.

6 Modern Methods of Analysis of Organic Components in the Atmosphere

In the past 10–15 years, the rapid development of analytical procedures has provided ample opportunities for the investigation of organic components of the atmosphere the content of which is much lower than $1 \times 10^{-6}\%$ by volume even in greatly polluted urban air. At present, various methods of determination of air composition are available to researchers. Active application of these methods has permitted great progress to be made in atmospheric chemistry. However, the quantity and quality of the available information are far from being sufficient for the solution of many important problems that have acquired paramount importance as a result of the ever-increasing influence of man on the environment. The lack of data is particularly felt when attempts are made to determine the consequences of the anthropogenic effect on the Earth's climate and the state of the ozone layer of the stratosphere and to simulate photochemical processes in the urban atmosphere.

This shortage of information is mainly due to very laborious procedures of investigations of atmospheric composition and to the high cost of complex analytical equipment. For example, the cost of such efficient instruments as a liquid chromatograph and a GC–MS system is many tens and even hundreds of thousands of dollars. Hence, the information on the organic components of the atmosphere is still insufficient and is often of fortuitous character.

Chromatographic methods have been used most widely both in the study of the composition of the atmosphere and in the investigation of photochemical reactions of organic compounds. This particularly concerns gas chromatography because of the relatively low cost of series instruments and wide possibilities of investigating volatile organic substances provided by this method The main problems related to the application of various types of chromatographic analysis arise in sample collection and its preparation for analysis. In contrast, procedures of analysis and processing of results usually do not involve great difficulties due to the modern level of automatization and the availability of selective detectors and packings of chromatographic columns.

In this chapter, information is provided on the principal methods of investigation of organic components present in the atmosphere in the gaseous and the condensed states. The purpose of this chapter is to help the reader dealing for the first time with the necessity for analyzing atmospheric air for organic compound content to make the correct choice of methods.

6.1   Chromatographic Methods

All the varieties of the chromatographic method of separation are based on the distribution of components of the mixture to be analyzed between two phases. The mixture is injected into the separating column and moves with the mobile phase along the layer of the stationary phase inmiscible with it. The rate of motion of each component depends on the interaction between the molecules and the stationary phase. The physico-chemical interactions ensuring the separation of compounds contained in the mixture can be different. They are the Van der Waals forces determining the adsorption of nonpolar molecules on nonpolar sorbents, induction interactions, the formation of hydrogen or ionic bonds, etc.

Depending on whether the mobile phase is a gas or a liquid, the process is called gas or liquid chromatography. Each of these variants has its own field of application, and their skilful combination enables us to obtain abundant information on the composition of the atmosphere. It is for this reason that almost all investigations of the composition of inorganic compounds and about half of investigations of organic gases in the atmosphere have been performed by chromatographic methods.

6.1.1   Gas Chromatographic Determination of Pollutants

The high sensitivity and unique possibilities of separation of extremely complex multicomponent mixtures have made gas chromatography an indispensable instrumental method for the investigation of minor gas components of the atmosphere, in particular, organic components. This method provided the greatest part of data available at present on the composition of air in background regions, rural areas and cities and on the sources of organic and some inorganic compounds.

The gas chromatographic (GC) determination of organic pollutants includes several main stages: sample collection, the preparation of the sample and its injection into the separating column, the GC separation of components and finally the processing of the results of analysis. Each of these stages can be carried out by different methods the choice of which depends on the final aim of the investigation and the availability of instruments. In the following sections, the main characteristics of the separating and detecting devices of modern chromatographs are given, and some variations of the most important stages of the determination of atmospheric pollutants, the collection of samples and their preparation for analysis, are discussed.

Gas chromatographic columns

At present, a wide range of chromatographic materials are being manufactured. These materials allow the separation of virtually any mixtures of volatile pollutants in atmospheric air. They include several hundred stationary phases differing in polarity and thermal stability, dozens of solid supports of the stationary phase, as well as porous materials for gas adsorption chromatography and chromatographic columns differing in size and manufactured from different materials.

In order to investigate in detail the composition of organic components of the atmosphere, high-performance capillary columns are required [366]. Metal or glass capillary columns 25–100 m in length with an inner diameter of 0.2–0.5 mm are usually employed for this purpose. Their walls are coated with a thin layer (0.1–0.25 μm) of various phases from nonpolar squalane to strongly polar Carbowax 20 M and FFAP. In the analysis of a wide range of organic components of the atmosphere, columns with stationary phases of low or moderate polarity (SE-30, OV-101 and DC-550 silicons, UCON LB-550 polypropylene glycol) are most often used. The application of these phases ensures better separation of hydrocarbons, the most abundant organic components of the atmosphere. A typical chromatogram of the components of atmospheric air separated on a capillary column is shown in Fig. 6.1. It can be seen that this chromatogram records the peaks of components the boiling temperatures of which range from − 12 °C to 176 °C, i.e. differ by almost 200 °C.

The ranges of boiling temperatures of volatile organic pollutants are very wide, and therefore in order to reduce the separation time simultaneously maintaining the performance level, temperature programming is used. In this case the choice of temperature regime is determined by both the properties of the

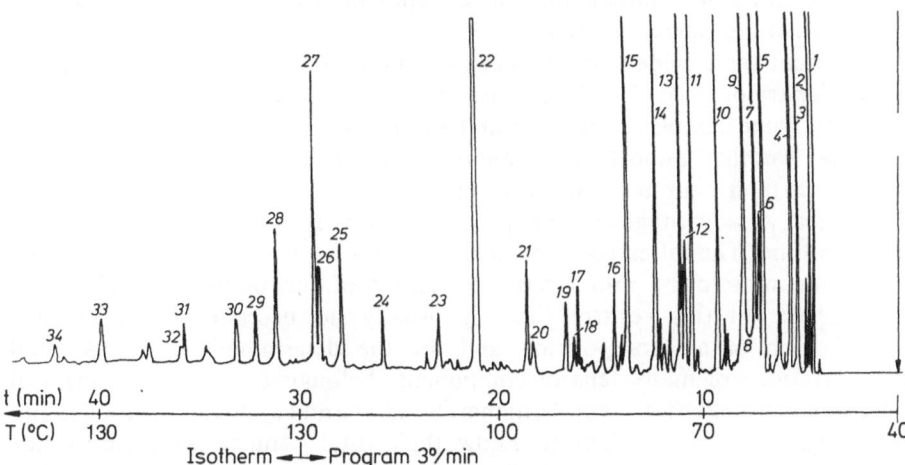

Fig. 6.1. Chromatogram of organic compounds of the atmosphere of Leningrad. Dinonylphtalate capillary column, 45 m × 0.25 mm ID, *1*- isobutane, *2*- *n*-butane, *3*- isopentane, *4*- *n*-pentane, *5*- 2-methylpentane, *6*-cyclopentane, *7*- 3-methylpentane, *8*- acetone, *9*-*n*-hexane, *10*- methylcyclopentane, *11*- 2-methylhexane, *12*-cyclohexane, *13*- 3-methylhexane, *14*- *n*-heptane, *15*- benzene, *16*-methylcyclohexane, *17*- 2-methylheptane, *18*- 4-methylheptane, *19*- 3-methylheptane, *20*- *trans*-1,4-dimethylcyclohexane, *21*- *n*-octane, *22*- toluene, *23*- trimethylcyclohexane, *24*- *n*-nonane, *25*- ethyl benzene, *26*- *p*-xylene, *27*- *m*-xylene, *28*- *o*-xylene, *29*- *n*-decane, *30*- *n*-propylbenzene, *31*- *m*-ethyltoluene, *32*- *p*-ethyltoluene, *33*-1,2,4-trimethylbenzene, *34*- *n*-undecane [96]

components to be analyzed and the temperature characteristics of the phase. For example, if C_2–C_4 hydrocarbons and some highly volatile methane derivatives (methyl chloride, CFM's) are among the compounds to be determined, the programming is carried out starting from low (sometimes even negative) temperatures with a long initial isothermic period. The presence of highly boiling compounds in the samples requires the application of thermally stable stationary phases that permit the decrease in the analysis time and in the background signal, which makes it difficult to carry out quantitative determinations. From this viewpoint, the most suitable columns are capillary columns with a bonded phase in which the molecules of the stationary phase are bonded by covalent bonds to the silanol groups on the inner surface of the glass columns. In this case not only the thermal stability of the phase but also the inertness of the glass surface increases because covalent bonds are formed with the participation of active silanol hydroxyl end groups.

The efficiency of analysis depends to a considerable extent on the column quality which, in turn, is dependent on the properties of the material from which the column is manufactured. Until recently, it seemed quite impossible to determine hydrocarbon concentrations on the level of 0.025 ppb typical of the background areas. However, this became possible by using columns made of fused quartz (fused silica). The surface of fused silica columns is very inert, and hence in these columns there is no spreading of the rear front of chromatographic zones even for polar components, which is characteristic of glass and, to an even greater extent, of metal columns.

In recent years, wide bore fused silica columns (0.75 mm) and a thick film of the stationary phase (1–5 μm) have been very widely used. In contrast to narrow columns, liquid samples of large volume can be injected into them without flow splitting. Another important advantage of fused silica columns over glass columns is their high mechanical strength.

A great disadvantage of all capillary columns is their high cost and long analysis time. The latter point makes it difficult to use capillary columns for systematic monitoring and operative control of the pollution level of the atmosphere. In this control there is usually no necessity for a detailed investigation of air composition and for the determination of individual concentrations of many tens of components belonging to the same class of organic compounds (for example, numerous alkane and cycloalkane isomers and homologues). It is sufficient to know their total content. In contrast, the information about photochemically active and toxic components: aromatic hydrocarbons, oxygen-, nitrogen- and sulfur-containing compounds and halo-carbons is of great interest. Most of these compounds are present in the atmosphere in ultra-low concentrations, and even when high-performance capillary columns are used, the peaks of these compounds are often overlapped and camouflaged by stronger alkane peaks. Express analyses and selective determination of the representatives of individual classes of organic compounds can be attained by using packed columns and specific detectors and by the application of the methods of selective concentration or the procedures of sample

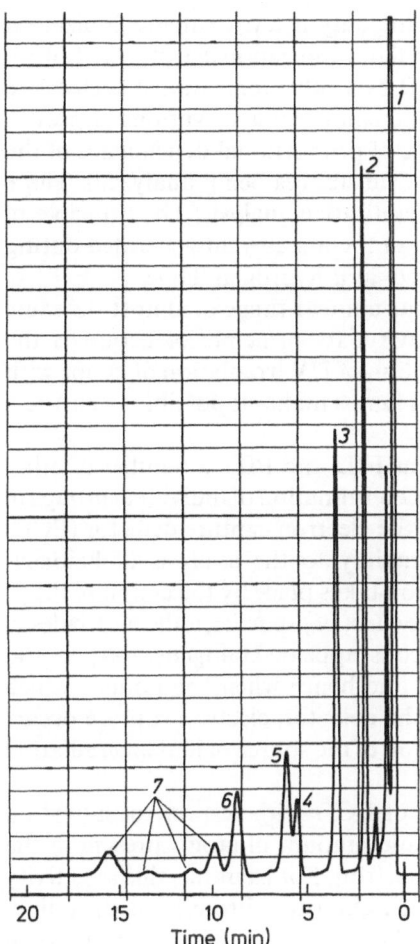

Fig. 6.2. Chromatogram of hydrocarbons in Leningrad air. *1*- alkanes and cycloalkanes C_4–C_{10}, *2*- benzene, *3*- toluene, *4*- ethyl benzene, *5*-*m,p*-xylene, *6*- *o*-xylene, *7*- C_9H_{12} alkylbenzenes. Column 3-m × 2 mm, 7.5% TCEP on Celite-545

Time (min)

preparation ensuring preliminary (precolumn) isolation of the group of compounds of interest from the entire amount of air pollutants.

Figure 6.2 shows an example of the determination of C_6–C_8 aromatic hydrocarbons on the background of C_4H_{10}–$C_{10}H_{22}$ alkanes and C_5H_{10}–$C_{10}H_{20}$ cycloalkanes. In this case selectivity is attained by the choice of column packing: 1,2,3–tris (cyanethoxy) propane, completely separating alkanes from aromatic hydrocarbons [99]. This is not the only example of group separation of air pollutants due to selectivity of column packing. Alkanes are very efficiently separated from alkenes and aromatic hydrocarbons on silver-containing zeolites [367].

GC Detectors

The success of analysis also depends to a considerable extent on the correct detector choice. Several tens of detectors for gas chromatographs are known.

However, modern instruments are sold with only a few of them. Ionization detectors are most widely used. Among them flame ionization detectors (FID) are employed most often. The advantages of FID over other ionization detectors are their high sensitivity with respect to organic compounds of various classes in combination with a wide linear range (range of proportional dependence of the detector signal on the concentration of the substances being analyzed). These properties make it possible to carry out quantitative analysis. The principle of action of FID is based on the measurement of the ion current generated during the chemoionization of combustion products in a hydrogen flame.

A photoionization detector (PID) exhibits a linear range of almost the same width as FID and an even higher sensitivity. Its principle of action is the ionization of molecules as a result of absorption of UV irradiation of lamps with an energy of 9.5 to 11.7 eV. A wide choice of lamps makes it possible to ensure a certain selectivity of analysis.

In addition to such universal detectors as FID and PID, a number of other element-selective detectors intended for the determination of individual groups of compounds are widely used. Among them, the electron capture detector (ECD) should be mentioned first. It is applied mainly to the analysis of halogen-containing compounds. In this case ionization takes place by the action of the β-irradiation of ^3H or ^{63}Ni isotopes. Under the effect of β-particles, the molecules of the carrier gas are ionized, and free electrons appear. Halogens, oxygen and nitrogen exhibit high affinity for electrons, and hence when compounds which contain the atoms of these elements enter the detector, electron capture occurs and the initial current formed due to the action of the source of ionizing radiation decreases.

It should be noted that the sensitivity of ECD with respect to individual compounds strongly depends on the type and amount of atoms present in the molecule which exhibit the affinity for the electron. For example, this sensitivity increases markedly in the series of fluorine-, chlorine-, bromine- and iodine-containing compounds. In the case of methyl chloride, CH_3Cl, the sensitivity of ECD is much lower than that of FID. However, this compound can be detected even in very low amounts if a reactor with KI is placed before the column. In this reactor, CH_3Cl is converted into CH_3I [368]. In this case the sensitivity of analysis increases by more than three orders of magnitude.

The thermoionic detector (TID) exhibits high selectivity with respect to nitrogen- and phosphorus-containing compounds. It is similar in design to the FID but differs from it in the ionization mechanism. Tablets with salts of alkaline metals ($CsBr$, $RbBr$, Rb_2SO_4, etc.) are used as the burner tip of the detector. When these salts enter the hydrogen flame, additional ion formation takes place during the combustion of nitrogen- and phosphorus-containing substances.

A flame photometric detector (FPD) is used in the analysis of admixtures of sulfur- and phosphorus-containing components. The combination of high selectivity and sensitivity makes FPD an invaluable tool for the determination of sulfur compounds on the background of hydrocarbons and many other compounds present in the atmosphere at much higher concentrations.

A highly sensitive helium ionization detector (HID) is much more seldom used in analytical practice, although it allows the determination of very small amounts of inorganic and organic compounds. This is evidently due to the imperfect design of the available detector models and the resulting instability of their operation. However, in future HID will doubtless be widely used for the analysis of various samples of the environment.

Sample collection and preparation for analysis

The entire number of organic components of the atmosphere may be tentatively divided into three main groups. The first group contains substances which are gaseous under normal conditions. They are C_1-C_4 light hydrocarbons and some of their derivatives. The second group comprises compounds liquid under normal conditions. The boiling temperatures of these substances are approximately in the range of those of liquid fuel for automobile transport. The third group contains hydrocarbons and their derivatives solid at room temperature. These compounds (for example, PAH) have low saturated vapor pressure and are readily sorbed on aerosol particles. The analysis of compounds in each of the above groups is carried out separately, and different techniques are used for sample collection.

There are several approaches to GC and other instrumental methods of impurity determination. One of them is the transport of analytical equipment directly to the location of investigations where automatic or manual sampling is carried out. Another method consists in air collection into specially prepared containers which are transported to central laboratories in which the air undergoes chromatography after preliminary operations. In the third method, the components to be determined are concentrated during sample collection, and subsequently the "concentrate" is transported to the laboratory and undergoes analysis after the usual preliminary treatment.

Each of these approaches has its field of application depending on both the range of compounds to be determined and economic considerations. These considerations drastically restrict the possibilities of carrying out investigations "on location" since chromatographic instrumentation is expensive. This approach is used in the organization of stationary control points (e.g. at stations of background monitoring) and in experiments carried out on ships, in airplanes and balloons, which often provide unique information.

Sample collection without previous concentration

The simplest method of sampling seems to be air collection into specially prepared containers. For this purpose, medical syringes or Tedlar bags up to several tens of liters in volume supplied with valves and stainless steel vessels 1 to 10 l in volume with a polished inner surface are used [62, 68, 243]. Metal vessels are often previously evacuated, and then sample collection may consist in the equalization of air pressure inside and outside the vessel. However, it is more convenient to prepare the sample for analysis if air is at an excess pressure. Hence, it is most often pumped by special pumps or introduced with a syringe through a self-sealing cap into the filled and sealed vessel.

The sensitivity of many detectors (e.g. ECD, PID and TID) is quite sufficient for a direct determination of many organic components of the atmosphere if only a few milliliters of air are injected into the column. However, since it is desirable to obtain more precise and detailed information about the composition of the atmosphere, it is necessary to apply intermediate concentration of the compounds contained in samples transported to the laboratory. This concentration is achieved by passing a fixed volume (up to several liters) of air through a cooled trap with an adsorbent or an inert filler packing. Figure 6.3 shows a typical scheme for gas lines used for injection into the chromatograph of a gas from a metal container under atmospheric pressure. This scheme has been used for the analysis of organic compounds in the air samples of tropical Brazilian forests [62]. Vessel (1) evacuated to the residual pressure of 10 torr induces the air flow. The volume of air passing through cooled concentrating column (7) from container (6) is determined by the change in pressure recorded with pressure gauge (2). The transfer of the concentrated components into the chromatographic column is carried out by thermal desorption after the switching of the carrier gas flow with the aid of a heated gas valve (5). Repeated sampling may be carried out after the gas is pumped out of vessel (1) through valve (3) at closed valve (4). The scheme is simplified if the air in the container is under excess pressure. Then the sample volume is determined with a flowmeter which replaces elements (1–3) in Fig. 6.3.

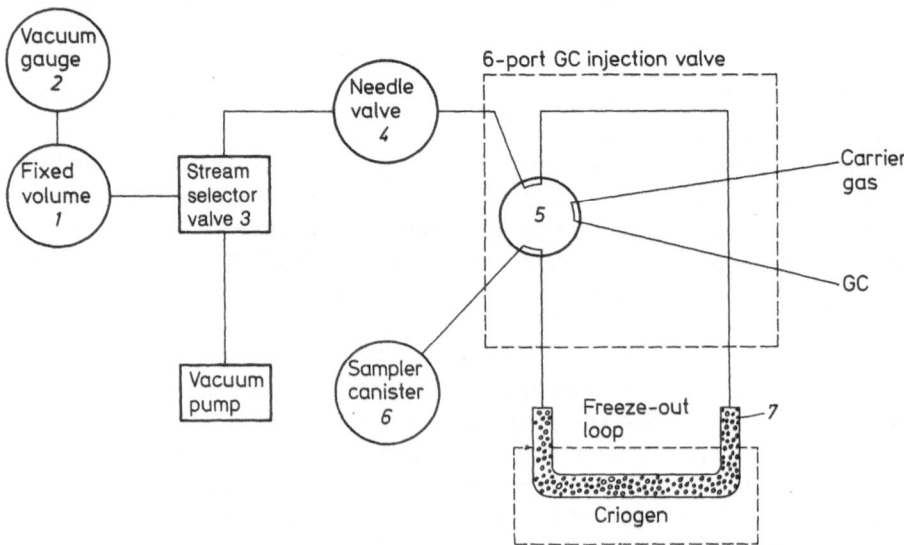

Fig. 6.3. Schematic of sample volume measurement system [62]

Hence, it should be noted that the above approach is characterized by the rapidity of sampling and the possibility of repeated analyses of the air transported to the laboratory (e.g. by using different gas chromatographic columns). Another important advantage is the possibility of carrying out very precise determinations of the sample volume with the aid of a pressure gauge. This is particularly important for sampling very rarefied air from aircraft. However, in practice this method requires considerable effort for the elimination of artefacts due to the evolution or, inversely, sorption of some components by the walls of vessels, connections and pumps.

It has been reported [68] that when air is collected in Tedlar bags, it becomes very polluted. Analysis of air collected in subtropical regions of the USA (Florida) has shown the presence of large amounts of acetaldehyde and acetone in all samples. Checking has shown that carbonyl compounds were evolved by the bag walls under the effect of light.

The precision of analysis can also be profoundly affected by sorption processes because we deal with the determination of microimpurities. Substance losses due to sorption can evidently be negligible in the case of gaseous and volatile nonpolar compounds, such as C_1–C_4 hydrocarbons or chlorofluorohydrocarbons. Nevertheless, they are very pronounced for polar and high boiling compounds: alcohols, aldehydes, higher alkanes and aromatic hydrocarbons. The losses of oxygen-containing compounds by sorption on the walls of steel containers have been reported in ref. [62]. The same authors have also pointed out that the quantitative determination of hydrocarbons with the number of carbon atoms more than eight (including $C_{10}H_{16}$ terpenes) is possible only when the air collected for analysis contains large amounts of water vapor. This fact indicates that hydrocarbons sorbed by the active centers on the container walls are competitively displaced by water molecules.

Another source of errors in this procedure of sample collection is the possibility of a change in air composition with time as a result of interaction between individual components. For example, it should be expected that unsaturated hydrocarbons (in particular, such reactive compounds as terpenes) should disappear, at least partly, in fast dark reactions with ozone constantly present in the open atmosphere.

Preliminary concentration

Preliminary concentration makes it possible to attain high analysis sensitivity with the aid of a high degree of sample concentration. Another important advantage of this method of sampling is the compactness of samples, the facility of their transport and the possibility of prolonged storage.

In analytical practice, low temperature (cryogenic) concentration, adsorption concentration (at the environmental temperature), chemosorption and equilibrium concentration are used most often.

Cryogenic concentration is performed by passing the air through cooled traps with various inert packings (glass wool, glass or teflon balls), adsorbents (silica gels, graphitized carbon black, and porous polymers) or packings for gas chromatographic columns with various stationary phases. Ice, solid carbon dioxide or liquefied gases are used for cooling. Concentrated compounds are transferred into the analytical column by heating the trap in the flow of the carrier gas. The defect of this method is the condensation of water vapor which often forms ice plugs making it difficult to collect samples and carry out gas chromatographic analysis. Hence, cryogenic concentration is usually applied only in those cases when other methods do not ensure quantitative trapping of the components in which we are interested, such as C_1-C_3 hydrocarbons and their volatile derivatives [59, 369].

Adsorption concentration is carried out at the temperature of the medium by using porous materials placed in sorption tubes. The materials used for concentration should meet several requirements [96, 101].

First, the adsorbent should trap compounds of different classes and different volatilities from as large volume of air as possible without their break through for attaining the required degree of concentration. At the same time it should not trap water vapor. Secondly, the sorbent surface should be sufficiently inert in order to ensure the invariable composition of the sorbate during sampling, storage and desorption. Since thermal desorption is the most convenient procedure for the transfer of concentrated impurities into the analytical column, the third requirement is the thermal stability of the sorbent (i.e. it should not decompose upon heating up to at least 300 °C).

At present, no sorbents meeting all these requirements are available. Hence, complex packing of sorption tubes is often used, the components of this packing complementing each other [96, 101]. Most often Tenax GC and Tenax T polymer sorbents and copolymers of styrene and divinyl benzene (chromosorb-102 and Porapack Q, etc.) are used for concentration. These copolymers have a very large surface area (250 to 750 m^2 g^{-1}) and adequately trap many volatile organic impurities of the atmosphere. A great defect of these copolymers is their low thermal stability (they begin to decompose at temperatures above 230 °C).

Carbonaceous sorbents, Carbotrap, Carbopack B and Carbochrom similar to them with a specific surface of 80 to 110 m^2 g^{-1}, Ambersorb XE and PSKT with a specific surface of 450 to 900 m^2 g^{-1} [88, 89, 96, 99] are also very popular. Carbonaceous materials are hydrophobic and exhibit very high thermal stability. The combination of sorbents with moderate (Carbochrom or Carbopack) and large (Ambersorb, PSKT) surface in a single tube makes it possible to concentrate and simultaneously determine the microconcentrations of a wide range of organic compounds: C_3-C_{12} hydrocarbons, alcohols, carbonyls and other impurities. It has been found, however, that during high temperature thermal desorption, catalytic transformations of many compounds. e.g. $C_{10}H_{16}$ terpenes and halocarbons [71, 99] can occur on the surface of carbonaceous sorbents. This is observed particularly often for the most widely used method of transfer of concentrated compounds into analytical columns: thermal desorption in a flow of an inert gas.

Secondary processes complicate to an even greater extent thermal desorption from the surface of such an excellent adsorbent as activated charcoal. Even relatively stable paraffin C_5-C_6 hydrocarbons undergo various transformations on the charcoal surface at temperatures as low as 200°C. In order to exclude artefacts, the components concentrated on activated charcoal are usually displaced by solvents [85], which leads to a decrease in analysis sensitivity [101]. Hence, the information about the possibility of using microwave irradiation for the removing of concentrated impurities from a trap with activated charcoal is of great interest [370]. According to the data of the author, desorption induced by microwave radiation for which activated charcoal is relatively transparent proceeds very rapidly and completely.

In recent years, another new method of sample preparation for analysis has been used. It consists of thermal desorption with the displacement of the sorbate with superheated water vapor instead of an inert gas. The advantages of this method are its rapidity, the possibility of the determination of unstable compounds without their decomposition at a relatively low desorption temperature and the preliminary (precolumn) separation of various classes of substances under analysis. This is attained because the carrier gas (water vapor) is condensed upon leaving the heated sorption tube and in combination with air displaced from sorbent pores forms a two-phase liquid-gas system. The distribution of desorbed compounds between these phases is determined by the ratio $C_L/C_G = K$ where C_L and C_G are the concentrations of compounds in the liquid and the gas, respectively, and K is the partition coefficient which for dilute solutions depends only on temperature. Hence, poorly soluble substances with low numerical K values are accumulated in the gas phase, and well soluble substances are accumulated in the aqueous phase. This method of sample preparation is particularly suitable for the analysis of low alcohols and carbonyl compounds on the background of predominant hydrocarbons.

Figure 6.4 shows the desorption scheme used by the author for the analysis of volatile organic components in urban air and in volcanic gases [99, 276, 371]. Superheated water vapor is injected into the sorption tube heated to 210 to 220°C packed with a complex absorber and connected with the vapor receiver of variable volume with the aid of a steel needle. This receiver is a 10 to 50 ml medical syringe. Water vapor and the air from sorbent pores push the syringe plunger,

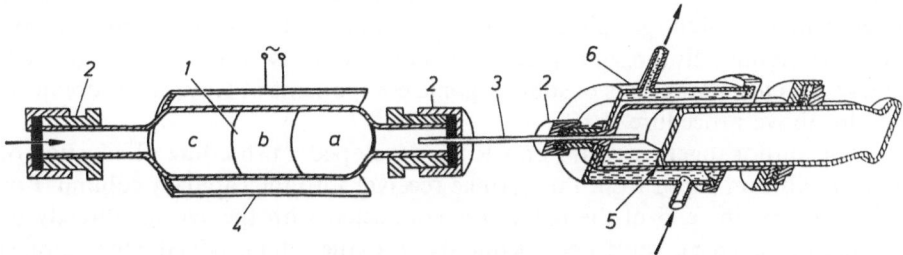

Fig. 6.4. Scheme for desorption by superheated water vapor in a variable volume device. *1-* sorption tube (*a-* Tenax GC, *b-* carbochrome, *c-*PSKT), *2-* union nut, *3-* steel needle, *4-* oven, *5-* syringe, *6-* poly(methyl methacrylate) thermostating jacket [371]

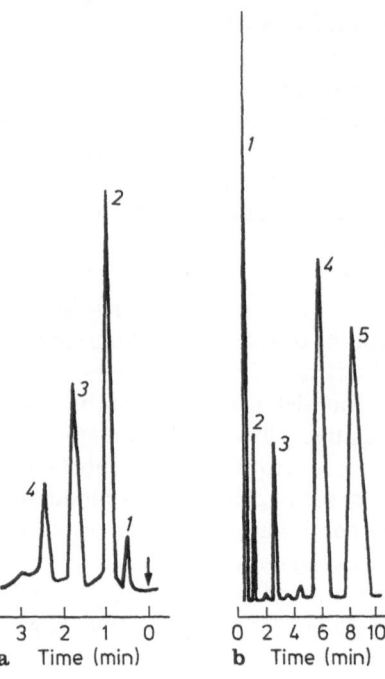

Fig. 6.5a, b. Chromatogram of some components in Habana air. (a) oxygenated compounds (*1*- methanol, *2*- acetaldehyde, *3*- ethanol, *4*- acetone). (b) halogenated compounds (*1*- air oxygen, *2*- CF_2Cl_2, *3*- $CFCl_3$, *4*- $CHCl_3$, *5*- CCl_4). The column (2 m × 3 mm) was packed with Separon BD sorbent [371]

and after desorption the pressure in the syringe becomes equal to the atmospheric pressure. After desorption is completed, the syringe receiver is disconnected from the sorption tube, and the gas is sampled for analysis through the rubber septum of the attachment point glued to the syringe cannula.

In this desorption method, it is possible to apply many effective procedures of headspace analysis (HSA), such as the replacement of the gas phase, the change in the values of the partition coefficient, etc. For example, the gas phase obtained in desorption is subjected to analysis for the content of hydrocarbons (with FID) and halocarbons (with ECD). The gas remaining in the syringe is pressed out and replaced by pure air, hot water from a thermostat is fed into the jacket, and a second analysis of the gas phase for the content of volatile alcohols and carbonyl compounds initially present in aqueous condensate is carried out. Figure 6.5 shows the chromatograms of some organic components of Havana air obtained by the above procedure.

The author together with Perez R. G. developed a procedure of injection of the equilibrium vapor from this syringe receiver into the capillary column. For this purpose, the end of the column is connected with the syringe directly or through a heated gas valve, and the initial part of the column coiled into a spiral is immersed in liquid nitrogen (Fig. 6.6). The gas flows from the syringe as a result of a pressure drop in the cooled part of the column. The gas volume in the syringe decreases but the pressure inside it remains virtually unchanged because of the

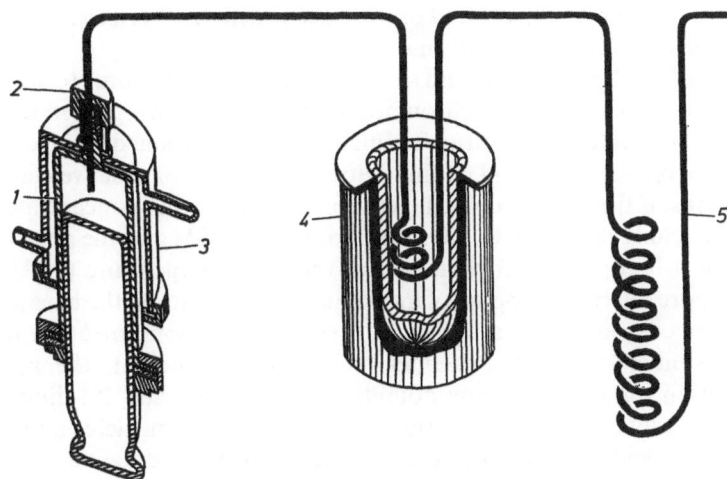

Fig. 6.6. Scheme for sampling equilibrium vapor phase into the capillary column. *1-* syringe cylinder, *2-* union nut, *3-* Plexiglass jacket, *4-* Dewar flask, *5-* capillary column

automatic compensation by the pressure of atmospheric air pushing the syringe plunger. This method differs in principle from pneumatic injection in which excess pressure is applied to the sealed vessel and then a part of the gas from the vessel to be analyzed is transferred to the analytical column operating at a lower pressure of the carrier gas [372]. In contrast, in this method the pressure in the column decreases continuously and therefore it may be called the "method of decompression injection".

The interphase distributions of compounds are also used in equilibrium concentration in which the air being analyzed passes through a suitable liquid until thermodynamically equilibrium distribution of impurities is established. Their concentrations in the liquid undergoing saturation are determined by the above ratio for the partition coefficient: the higher the numerical value of K for a compound in the system, the greater degree of its concentration in the liquid can be attained during sampling [372].

The advantage offered by the method of equilibrium concentration is that it does not require precise measurements of the air volume. Moreover, the sample can be subjected to repeated analyses, and since the stage of thermal desorption in the preparation for analysis is absent, the losses caused by incomplete desorption, decomposition and other chemical transformations of unstable compounds are excluded.

The trapping liquid may be chosen in such a manner that selective accumulation of certain groups of organic compounds takes place. For example, the most suitable solvent for concentrating lower alcohols, carbonyl compounds and amines is water. Concentrated or 80%-acetic acid has been successfully used for trapping aromatic hydrocarbons.

Both the liquid concentrate and the gas phase allowed to come to equilibrium with it can be subjected to gas chromatographic analysis. Moreover, the change in the numerical values of partition coefficients is often attained by increasing the temperature of the system and adding salting-out agents or other reagents to the system. Thus, if the partition coefficient for benzene in the concentrated acetic acid-air system is 780 at 20 °C, after the addition of KOH and the neutralization of the acid it becomes lower than four at the same temperature [372].

Even more selective trapping of individual components on the background of many other compounds is attained in chemisorption concentration based on using a number of specific chemical reactions. At present, the method of concentration of carbonyl compounds by reaction with 2,4-dinitrophenyl hydrazine is widely used. The reaction is catalyzed by strong acids and proceeds at a high rate with the formation of 2,4-dinitrophenyl hydrazones:

$$RR'C{=}O + NH_2NHC_6H_3(NO_2)_2 \longrightarrow RR'C{=}NNC_6H_3(NO_2)_2 + H_2O$$

After extraction with organic solvents, the resulting hydrazones are subjected to gas chromatographic analysis with the aid of a TID or ECD detector sensitive to nitrogen atoms [15, 16].

In our laboratory, film sorption tubes are employed for the concentration of carbonyl compounds in the form of 2,4-dinitrophenyl hydrazones. Film sorbents developed at the Main Geophysical Observatory of the USSR are glass tubes containing a layer of grains of a non-porous support (glass) fixed between two perforated partitions (Fig. 6.7). The glass grains are coated with a film of non-drying reagent solution. Ethylene glycol or glycerol are used as admixtures ensuring the non-drying of reagent solutions. The use of these sorbents has many advantages: the simplicity of preparation for sample collection, good reproducibility of sorbent properties, the facility of attaining complete desorption of concentrated compounds and the compactness of tubes (usual tube size is $170 \times 12\,mm$).

We used tubes in which glass grains were wetted with a saturated solution of 2,4-dinitrophenyl hydrazone in sulfuric acid [373]. The resulting hydrazones were dissolved in ethyl acetate and subjected to gas chromatographic analysis by using FID or ECD. With a sample volume of 91, minimum detectable concentrations of formaldehyde and acetaldehyde were $20\,\mu g\,m^{-3}$ in analysis with FID and $0.2\,\mu g\,m^{-3}$ in analysis with ECD. The total experimental error was $\pm 15\%$.

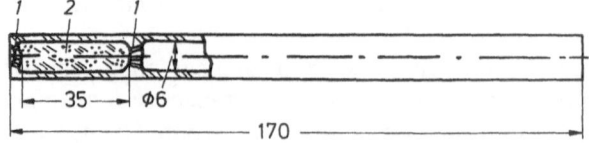

Fig. 6.7. Film sorption tube; *1*- perforated partitions, *2*- sorbent (glass grains covered with a non-drying reagent solution)

Chemisorption concentration in combination with headspace analysis has been successfully used for the determination of amine content in the atmospheric air [374]. For trapping the amines, air is passed through a tube packed with glass grains with a sulfuric acid solution fixed on them. The remaining amine salts are washed off with water, and the resulting concentrate is treated with potassium hydroxide in a sealed vessel. The amines pass into the gas phase which is subjected to chromatographic analysis.

The alkaline components are trapped in sorption tubes with an acid, whereas the method of chemisorption in tubes containing alkalis is quite applicable to the concentration of volatile C_2–C_6 acids and phenols. For this purpose, glass grains can be treated with a weak aqueous KOH or NaOH solution.

Suitable reagents can be selected for the concentration of many other hydrocarbon derivatives. For example, mercaptans are quantitatively bonded by salts of bivalent mercury.

$$2RSH + (CH_3COO)_2Hg \longrightarrow (RS)_2Hg + 2CH_3COOH$$

Nanogram amounts of hydrogen sulfide, methyl mercaptan and dimethyl sulfide can be detected with the aid of FPD after chemisorption concentration in quartz sorption tubes packed with ultrathin ($150\,\mu$m) gold fibers [271].

In conclusion, we should briefly deal with the methods of collecting samples of atmospheric aerosols and their preparation for gas chromatographic determination of organic compound content. The aerosol particles are trapped with the aid of filters or instruments for inertional precipitation (impactors). Fine fiber glass and polymer filters are most often used. For analyzing the organic components of aerosols, very large volumes of air (from several hundred to several thousand cubic meters) are usually passed through the filters at a rate of 20–$100\,m^3\,h^{-1}$ and higher.

The stage of sample collections introduces the largest error into the results of analysis because of the error in the determination of sample volume and the loss in organic components during the aspiration of air. This loss is due to the volatilization of a part of the organic compounds during the passage of large air volumes through filters and impactors and their interaction with the chemically active components of the atmosphere: ozone, nitrogen oxides, sulfur dioxide and acids [375]. Thus, polynuclear aromatic hydrocarbons in aerosols collected on filters can undergo partial nitration, sulfonation and structural degradation due to the effect of these agents and sunlight. Hence, current procedures of sample collection do not adequately reflect the composition of organic components in aerosols and should be improved.

A detailed study of the chemical composition of atmospheric aerosols which contain hundreds of organic components of various classes is a very complex problem. Most often, the preliminary separation of the entire amounts of organic compounds by the methods of column liquid chromatography and thin-layer chromatography into separate fractions is used with subsequent gas chromatographic analysis of each fraction. A typical scheme for aerosol preparation for analysis is shown in Fig. 6.8 [376].

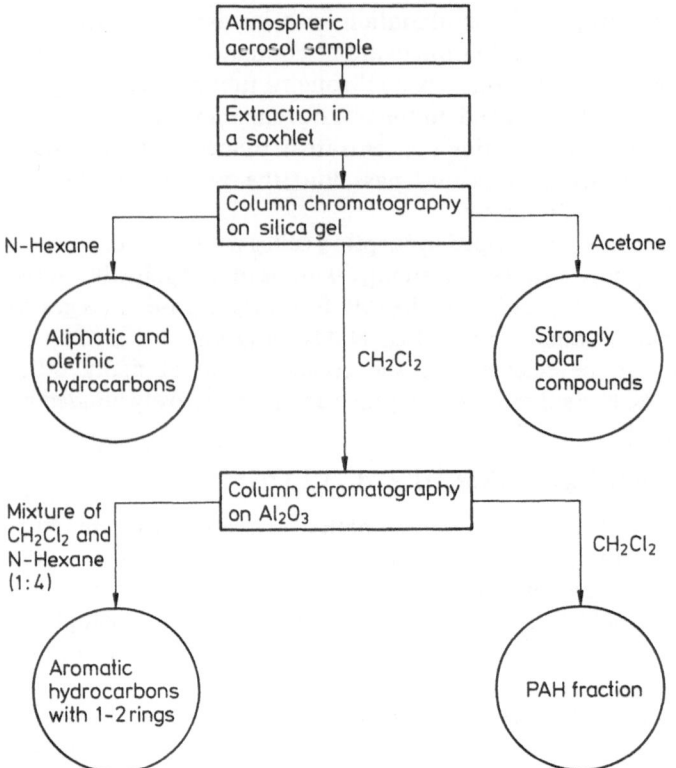

Fig. 6.8. A typical scheme of fractionation of aerosol organic substituents [376]

Organic components are isolated by many hours extraction on filters in a soxhlet. For a more complete extraction, glass filters are ground and treated with solvents in ultrasonic baths. Benzene, methylene chloride and methanol are usually employed as solvents. The extract is centrifuged, evaporated to dryness, dissolved again in a smaller amount of the solvent and separated into fractions by passing the solution through a short column packed with silica gel. Fractionation is achieved by using eluents characterized by polarity. Nonpolar compounds (alkanes, cycloalkanes and alkenes) are displaced from the column with hexane or cyclohexane, weakly polar compounds and PAH are eluted with methylene chloride and strongly polar compounds are displaced with acetone. The fraction of weakly polar compounds and PAH is again separated on a column packed with active aluminium oxide or Sephadex. A mixture of methylene chloride and hexane displaces aliphatic compounds and low molecular weight aromatic hydrocarbons, benzene and naphthalene derivatives, whereas pure CH_2Cl_2 elutes PAH. After the evaporation of the solutions to dryness, the residue is again dissolved in a minimum amount of an appropriate solvent and chromatographed on capillary columns.

In the analysis of organic compounds of the atmospheric aerosol, it is evidently possible to use a much simpler thermal desorption method of component transfer from filters to analytical columns. The authors of ref. [377] have proposed placing a part of the teflon filter into the desorption tube heated for 5 min to 350 °C in a helium flow and connected to the capillary column. In this case, the detection limit is 0.01 pg m^{-3}, i.e. the sensitivity is higher by about two orders of magnitude than that attained with liquid extraction. The main difficulty consists in the deciphering of a very complex chromatogram containing the peaks of many hundred components. It is possible to simplify to a certain extent the problem of this deciphering by using element-selective detector [378], e.g. ECD for the separation of peaks of non-volatile polychlorine-containing compounds (polychlorinated biphenyls, dibenzofurans and dibenzodioxines), TID for the recognition of peaks of nitroderivatives of PAH and phosphorus-containing components, etc.

Recently very interesting information has appeared concerning the investigation of the possibility of using supercritical fluid extraction (SFE) for the extraction of adsorbed PAH [379]. It has been established that SFE proceeds much more rapidly (about 30 min) and more efficiently than extraction in a soxhlet. It is recommended that 2-methyl propane be used as an eluent for PAH and methanol with the addition of 20% CO_2 for high-boiling polar compounds. Extraction is carried out at pressures of about 40 MPa and temperatures below 235 °C. One may hope that the application of these new approaches will make it possible to increase drastically analysis sensitivity with a simultaneous decrease in the volume of the air sample and thus enhance reliability of information about the organic components of the atmospheric aerosol.

6.1.2 Liquid Chromatography in the Analytical Chemistry of the Atmosphere

For a long time, liquid chromatography, a historically earlier variation of the chromatographic method, has not been as widely used for analysis of atmospheric air and other natural environments as gas chromatography. The restricting factors were mainly long analysis time and the absence of detecting devices of continuous operation. However, liquid chromatography exhibits many characteristics which make it more advantageous than gas chromatography. The latter is applicable to the separation of mixtures of compounds with a considerable value of vapor pressure at the analysis temperature but not all of them are stable under these conditions and often undergo thermal degradation. Liquid chromatography does not require the transition of the components being analyzed into the vapor phase and is carried out at relatively low temperatures at which the decomposition of molecules of high boiling and thermally unstable compounds does not occur. Another advantage is the very high efficiency of this method resulting from a very high degree of homogeneity of finely dispersed sorbent used as the stationary phase or its support.

The success of liquid chromatography observed from the beginning of the

1970s is to a considerable extent related to the appearance of surface-porous sorbents which can withstand high pressure and in which the rate of the mobile phase has little effect on the separation efficiency. It became possible to design instruments operating at very high pressures (20 MPa and higher), which resulted in a decrease in analysis time by several tens of times. The method of liquid chromatography carried out at high pressure is called high-performance liquid chromatography (HPLC).

Devices based on various principles are used as detectors. Differential refractometer has been widely used as a universal detector. It is based on the continuous measurement of the refractive index of the liquid eluted from the column and its comparision with that of the solvent.

Spectral detectors are used in the analysis of substances absorbing in a certain spectral region ranging from ultraviolet to infrared region. Fluorimetric and polarographic detectors ensuring the possibility of analysis of the components at concentrations of 10^{-2} to $10^{-4} \mu g\, ml^{-1}$ are also very sensitive.

Ionization detectors also used in liquid chromatography are similar in their principle of action to the above-mentioned flame-ionization, electron-capture and other detectors. All of them comprise a group of the so-called transport-ionization detectors.

At present, surface-porous sorbents are being replaced with finer dispersed volume-porous sorbents with particle diameter less than 10 μm. Silica gels with a chemically modified surface are widely used. Modification consists in the grafting of various groups onto silanol groups on the sorbent surface: trimethylsilane, octyldimethylsilyl, octadecyldimethylsilyl, diphenylmethylsilyl groups, etc. Silica gels with octadecyldimethylsilyl groups (μ-bondapack C_{18}, Zorbax ODS C_{18}, Microsorb RP-C_{18} and Supelcosil LC-18) are often used for the separation of organic components of atmospheric air.

The columns of liquid chromatographs usually operate at room temperature, and only for the solution of some specific problems (e.g. polymer analysis), is it necessary to raise the temperature. Hence, in HPLC, temperature programming has been replaced by gradient elution during which the properties (most often, polarity) of the solvents gradually change or by stepwise elution in which a fast transition from one isocratic regime to the other takes place.

As an example of the application of HPLC to the analytical chemistry of the atmosphere one can mention the investigation of the composition of polynuclear aromatic hydrocarbons in the atmospheric aerosol. In this case HPLC is used either as an independent method of analysis or for the preliminary isolation of the PAH fraction from the concentrates obtained by extraction of filters followed by their subsequent detailed gas chromatographic investigation on capillary columns. Water-methanol (1:9) or water-acetonitrile mixtures are generally used as the mobile phase under the conditions of isocratic or gradient elution.

PAH are recorded with the aid of an UV detector in the spectral range of 280–365 nm or a more sensitive and selective fluorimetric detector in which excitation occurs with light at 280–360 nm wavelengths and luminescence recording takes place in the 370–600 nm range. The fluorimetric detector makes it possible to detect PAH contained in the amount of 0.1 ng ml^{-1}.

HPLC is used for the determination of the content of not only high-boiling components of the atmosphere but also of a number of highly volatile organic compounds. In the latter case they are transformed into various derivatives. As we have already mentioned, some impurities are concentrated by chemisorption. The resulting compounds can be analyzed by HPLC. For example, 2,4-dinitrophenyl hydrazones of aldehydes and ketones show strong absorption bands in the near UV spectral region and are recorded by an UV detector at 254 and 360 nm [104, 105, 108, 109]. This method is very sensitive: the lowest limit of formaldehyde detection can attain 0.01 ppb [105]. Hydrazones are separated on short columns packed with octadecyl sorbents under isocratic conditions of selection with water-methanol or water-acetonitrile (4:6) solutions.

In conclusion, another variant of liquid chromatography should be mentioned in which an ion exchanger is the stationary phase in the column. This allows the efficient separation of cations and anions. Macroporous ion-exchange XAD resins bearing amino groups on the surface, Dowex type resins and silica gels with covalently bonded quaternized amino groups or alkylsylphonyl groups may be used as column packings. In the case of ionic chromatography, aqueous buffer solutions or solutions of organic electrolytes serve as the mobile phase. Detection is performed with the aid of either ultraviolet, conductometric or ion-selective detectors.

Ionic chromatography may be successfully applied to the detection of amines after their chemisorption concentration in the form of salts and to the analysis of organic acids [126] and aldehydes after their oxidation.

6.1.3   Hyphenated Systems

The methods and instrumentation combining chromatography as one of the most efficient methods of separation of complex multicomponent mixtures with various spectral detectors having unique possibilities of establishing the structure of organic compounds have been used in the analytical chemistry of the atmosphere for more than 15 years. The attempts of developing combined methods of analysis have already been undertaken previously. In particular, the trapping of components at the outlet of the chromatographic column and the subsequent recording of IR or mass spectra with appropriate instruments operating under autonomous conditions have been described. However, it became possible to use in practice all the advantages of the combination of these fundamentally different methods only after the development of a single analytical system containing three principal elements: a chromatograph for the separation of the components, a spectrometer serving as a selective detector and a computer employed for controlling the entire system, for data processing and for the presentation of results of analysis.

The combination of gas chromatography and mass spectrometry (GC-MS) has become most widely used for the investigation of the composition of the atmosphere. With the aid of GC-MS, the presence of hundreds of organic compounds in the open atmosphere was established as early as in the 1970s, and the information about their main sources was obtained [101].

At present, the identification of components is characterized by a complex approach with the application of both purely chromatographic data (retention parameters of components in chromatographic columns) and mass-spectrometric information proper. In the GC-MS analysis of air, all the range of procedures of chromatographic analysis is now being used: preliminary concentration of pollutants, their chemical modification, precolumn separation, etc.

For a relatively long time, GC-MS was less sensitive than gas chromatography. This was due to the fact that a much larger amount of the substance is required for the recording of a complete mass spectrum suitable to identification than for its simple detection. However, the development of procedures of mass-fragmentography (selective detection of individual ions) has made it possible to decrease the detection threshold to 10^{-9} to 10^{-12} g, i.e. to exceed the sensitivity of some selective gas chromatographic detectors. Mass-fragmentography provides a considerable gain in sensitivity because the detection efficiency of selected ions is higher than that for the recording of the complete spectrum.

The use of a computer permits simultaneous recording of several ions (in some models of up to 20 ions). For example, the Hewlett-Packard HP-5970 A gas chromatographic detector consisting of a mass-spectrometric analyzer and a control block makes it possible to register simultaneously six different ions and to record the so-called mass chromatograms. This technique is particularly convenient for the search of certain components in complex mixtures with an unknown composition (in this case selective detection is carried out by using the most characteristic peaks in the spectra of the components).

In recent years, great progress has also been made in the development of analytical instrumentation combining HPLC and mass spectrometry (HPLC-MS). In this case the main difficulty was the development of a system for efficient removal of the solvent. At present these difficulties have mainly been overcome, and various companies manufacture series instruments for HPLC-MS.

The instruments for GC-MS are among the most expensive analytical equipment. However, their application ensures the qualitative and quantitative characterization of complex samples with an unknown composition and requires three or four chromatographs equipped with various detectors and columns. Economic calculations show that capital outlays (if the time needed for complete analysis and the number of attending staff are taken into account) are even smaller than by using only gas and liquid chromatographs.

A combination of gas chromatography and optical spectroscopy is also very promising for the analysis of environment samples. IR spectroscopy is a source of valuable information on the structure of organic compounds but its sensitivity is low. The combination of an IR spectrometer and a gas chromatograph has become possible only with the application of a computer. At present, a number of companies produce instruments containing a Fourier spectrometer as detector. Its principle of action consists of the two-stage recording of the spectrum: first, the interferogram of radiation passing through the sample is obtained with the aid of a Michelson interferometer, and subsequently with its aid the spectrum is calculated by using the Fourier transform.

These instruments consists of a gas chromatograph, an interferometer, an amplifier, an analogue-digit transformer and a computer. The main advantage of Fourier spectroscopy over other types of optical spectroscopy is the possibility of simultaneous recording of the entire spectrum in a very short time and with a higher resolution. The speed and sensitivity of these instruments are sufficient for the identification of organic compounds in operations with capillary columns at the detection level of a few fractions of a microgram [380]. The combination of a Hewlett-Packard 5600 GC gas chromatograph and an FT-IR Nicolet 60SX spectrometer may be reported as an example of a successful GC-FT-IR system.

A combination of gas chromatography with atomic absorption spectroscopy (GC-AAS) is also widely used for the analysis of organometal compounds. AAS is a very selective and sensitive method. The GC-AAS analytical instruments have been employed for the detection in air of micro-impurities of selenium, arsenic, lead, mercury, manganese and other toxic metals [185–187, 381]. The procedure usually involves the concentration of organometal compounds on porous sorbents, subsequent thermal desorption directly into the gas chromatographic column or into a trap containing an organic solvent. Atomization after the separation on the column is carried out either directly in the AAS flame or in a pyrolyzer at the column outlet. The detection limits is on the level of fractions of a nanogram in the sample. GC-AAS may also be used in the analysis of metals in the form of their volatile complexes with organic ligands and in the form of volatile carbonyls.

6.2 Spectroscopic Methods of Investigation

At present, the methods including air sampling and subsequent laboratory analysis are the main tools for the study of the chemical composition of the atmosphere. As already mentioned, the extent of information available with their aid does not meet modern requirements. Moreover, the results of analysis indicate the chemical composition of air at the location of sampling, whereas the information about the content and distribution of components throughout the atmosphere and in its different layers is also of great interest. Under ideal conditions, this information should be sufficient for the plotting of the time-dimension dependence of the content of organic compounds of interest. It is possible to plot this dependence by using some spectral methods: absorption and emission spectroscopy.

In the investigations of the atmospheric composition, absorption spectroscopy uses the Sun as the radiation source. Solar radiation passing through the atmosphere is recorded with a spectrometer. The analysis of lines in the absorption spectrum enables us to draw conclusions about the presence and the quantity of a component. Measurements may be carried out over a very wide range from the ultraviolet to the far infrared region. As far back as 1948 it was established that methane is present throughout the troposphere, and later the presence of formaldehyde was also detected.

Emission spectroscopy uses an artificial source of excitation which transforms the molecules of the compound being analyzed to the states with a higher energy. Inverse transition is accompanied by the irradiation in the ultraviolet, visible or infrared spectral region. Lasers have been widely used as these sources.

All these methods are called remote sensing because they provide the characterization of the chemical composition not at a single point but, rather, averaged over the path from tens of meters to hundreds of kilometers depending on the instrumentation and the procedures employed. Hence, they make it possible to obtain data on the space-dimension changes in the chemical composition of atmospheric air over a large territory and, if artificial satellites are used, even on the global scale.

In the UV and IR regions, the absorption spectra of many microcomponents of the atmosphere overlap each other as well as the absorption bands of carbon dioxide and water molecules. Consequently, the resolution of the instrument should be very high. The parameters according to which a compound is identified and quantitatively determined are the absorption coefficients and the elements of the fine structure of spectral lines forming the absorption bands. However, these parameters depend not only on the concentration of the compound but also on its distribution in the investigated layer and the characteristics of the state of the atmosphere: temperature, pressure and the content of light scattering aerosol particles. Therefore, although the principle of remote sensing of the atmosphere by using the Sun as the radiation source seems very simple, in practice this method involves considerable difficulties. Hence, at present one can speak about the prospects of this method in the near future rather than about its present success.

Dianov-Klokov B.I. et al. [18, 19, 20] have used both stationary spectral instruments and those mounted on mobile platforms with the resolution of 0.3 cm^{-1} for the investigation of the v_3 methane absorption band (about 3.3. μm). Quantitative determination has been carried out according to the areas inside the contour of the P(2) line of the v_3 band.

The measurement of the content of other organic components from absorption spectra at the present levels of their concentrations in the upper troposphere and the stratosphere requires higher resolution. For example, in order to determine formaldehyde content, an instrument with the resolution of 0.1 to 0.01 cm^{-1} is needed.

Recently, reports have appeared that periodic or systematic use of these instruments for the investigations of microimpurities in the atmosphere has already started [48, 64, 65]. The absorption spectra in the 8 to 13 μm range recorded at sunset and sunrise in a mountain observatory in the Pyrenees have been used for the calculation of the integral concentration of CFC's. In the same spectral range, solar radiation has been recorded since 1981 with a Fourier-spectrometer with a resolution of about 0.005 cm^{-1} at the Kitt Peak Observatory

(Arizona, USA) [382]. In this observatory, apart from CFC's, the integral ethane concentrations has been determined [65]. Ethane concentration measurements from balloons and aircraft at altitudes of 12 to 33 km using spectrometers with 0.06 to 0.02 cm^{-1} resolution have been carried out by the authors of ref. [64].

In high latitudes, the conditions for spectroscopic investigations (low height of the Sun above the horizon, low humidity and low general atmospheric pollution) are particularly favorable. Until recently, only single short-term observations were carried out in polar regions (e.g. on the Zhokhov Island in the East-Siberian sea [383]. At present, systematic observations using a high resolution Fourier spectrometer have started at the Japanese Antarctic station Syowa.

Investigations with the aid of satellites have also begun. For these investigations, instruments with high and super-high resolution have been developed. They are intended for the determination of the content of many important microcomponents of the atmosphere including the molecules of organic substances: methane, formaldehyde, methyl chloride, chlorofluorohydrocarbons.

The development of tuning lasers covering a wide spectral range from the ultraviolet to the far infrared region is a great incentive to extensive application of the laser technique to remote sensing of the atmosphere. These new laser methods employ various effects: the absorption of transmitted radiation, Raman scattering and fluorescence. In the near future, a possibility will probably appear of applying the laser technique to remote control of the content of some organic compounds in satellite experiments.

Hence, a tendency towards the applications of unique expensive instrumentation to the investigations is observed but it is justified by the possibility of obtaining with its aid extensive information on the global distribution of microcomponents of the atmosphere. So far information is available about successful tests of cheaper laser instruments in land experiments with the path length of about 1 km. According to this information, the detection limit of a number of organic compounds is 2–3 ppb. The development of cheap mass instrumentation of this kind would solve the problem of operative automatic control of the state of the urban atmosphere and of the activity of individual stationary sources of atmospheric pollution.

In conclusion it should be noted that spectral methods are begining to be widely used not only in the remote sensing of the atmosphere but also in the analysis of the component content at a location. For this purpose instruments are used with multiple-pass optical cells in which the absorption of IR or UV radiation by air components is determined. The samples transported into the laboratory can undergo analysis or else the air under analysis is continuously bubbled through the flow cell of the spectrometer mounted on a moving platform.

Luminescent analysis is one of the most sensitive methods of emission spectroscopy. Luminescent spectroscopy with the application of the Spol' skii effect has been most widely used in the analysis of PAH [384]. The Spol' skii effect is the formation of electron-vibrational spectra consisting of series of narrow spectral lines (quasi-linear spectra) under the conditions when PAH molecules are rigidly fixed in the solvent. This is accomplished by freezing PAH solutions in alkanes (usually in *n*-octane) at a temperature of liquid nitrogen or helium. The quasi-linear spectra obtained are strictly individual for each organic compound. Quantitative analysis is carried out from the concentration dependence of the intensities of characteristic bands.

A considerable advantage of this method is that complete fractionation is not needed. The detection limit of PAH is on the level of $10 \, \text{ng ml}^{-1}$.

References

1. Valley SL (ed) (1965) Handbook of Geophysics and Space Environment. McGraw-Hill, New York
2. Folsome CE (1979) The Origin of Life. A Warm Little Pond. WH Freeman, San Francisco
3. Awramik SM, Cloud P, Curtis CD, Folinsbee RE, Holland HD, Jenkyns HC, Langridge J, Lerman A, Miller SL, Nissenbaum A, Veizer J (1982) Biogeochemical Evolution of the Ocean-Atmosphere System. State of the Art Report. In: Holland HD, Schidlowski M (eds) Mineral Deposits and the Evolution of the Biosphere, Dahlem Konferenzen. Springer, Berlin Heidelberg New York, pp 309–320
4a. Markhinin EK (1985) Volcanism. Nedra, Moscow
4b. Sagan C (1965) Primary synthesis of nucleoside phosphates under the influence of UV radiation. In: Fox SW (ed) The origin of prebiological systems and of their molecular matrices. Academic Press, New York
4c. Garrels RM (1975) Circulation of carbon, oxygen and sulfur during geological time. Nauka, Moscow
5. Budyko MI, Ronov AB, Yanshin AL (1985) History of the Atmosphere. Gidrometeoizdat, Leningrad
6. Margulis L, Lovelock JE (1974) Icarus 21:471
7. Neftel A, Moor E, Oeschger H, Stauffer B (1985) Nature 315:45
8. Vager BG, Turchanovich IE (1987) Meteorol. i Gidrol. N 9:39
9. Wang WC, Pinto JP, Yung YL (1980) J. Atmos. Sci. 37:333
10. Kagann RH, Elkins JW, Sams RL (1983) J. Geophys. Res. C88:1427
11. Dickinson RE, Cicerone RJ (1986) Nature 319:109
12. Cavanagh LA, Schadt CF, Robinson E (1969) Environ. Sci. Technol. 3:251
13. Khalil MAK, Rasmussen RA (1983) J. Geophys. Res. C88:5131
14. Ehhalt DH, Heidt LE (1978) J. Geophys. Res. 78:5265
15. Ehhalt DH (1974) Tellus 26:58
16. Aikin AC, Gallagher CC, Spicer CW, Holdren MW (1987) J. Geophys. Res. D92:3135
17a. Reihle HG, Condon EP (1979) Geophys. Res. Lett. 6:949
17b. Migeotte M (1948) Phys. Rev. 73:519
18. Dianov-Klockov VI, Lukshin VV, Matveeva OA, Sklyarenko IYa (1977) Izvestiya AN SSSR. FAO 13:529
19. Dvoryashina EV, Dianov-Klockov VI, Fokeeva EV, Yurganov LN (1982) Izvestiya AN SSSR. FAO 18:46
20. Fabian P, Borchers R, Weiler HG (1979) J. Geophys. Res. C84:3142
21. Newell RE, Condon EP, Reichle HG (1981) J. Geophys; Res. C86:9833

22. Khalil MAK, Rasmussen RA (1984) Antarctic J. US 19:204
23. Blake DR, Rowland FS (1986) J. Atmos. Chem. 4:43
24. Rasmussen RA, Khalil MAK (1981) J. Geophys. Res. C86:9826
25. Blake DR, Mayer EW, Tyler SC, Mahide Y, Montague DC, Rowland FS (1982) Geophys. Res. Lett. 9:477
26. Fraser PJ, Hyson P, Rasmussen RA, Crawford AJ, Khalil MAK (1986) J. Atmos. Chem. 4:3
27. Larson RE, Lamontagne RA, Wilkniss PE (1972) Nature 240:345
28. Ehhalt DH (1978) Tellus 30:169
29. Malkov IP, Dianov-Klockov WI, Lukshin VV (1980) Izvestiya AN SSSR. FAO 16:763
30. Coffer WJ (1982) J. Geophys. Res. C87:7201
31. Heidt LE, Krasnec JP, Lueb RA, Pollock WH, Henry BE, Crutzen PJ (1980) J. Geophys. Res. C80:7329
32. Behar JV, Zafonte L, Cameron RE, Morelli FA (1972) Antarct. J. US. 7:94
33. Seiler W, Müller F, Oeser H (1978) Pure Appl. Geophys. 116:554
34. Fabian P, Borchers R, Flentje G, Matthews WA, Seiler W, Giehl H, Bunse K (1981) J. Geophys. Res. C86:5179
35. Bush YA, Schmeltekopf AL, Fensenfeld FS, Albritton DL, McAfee JR, Golden PD, Ferguson EE (1978) Geophys. Res. Lett. 5:1027
36. Schmidt M, Borchers R, Fabian P, Flentje G, Matthews WA, Szabo A, Lal S (1984) J. Atmos. Chem. 2:133
37. Khalil MAK, Rasmussen RA (1987) Atmos. Environ. 21:2445
38. Stephens ER (1985) J. Geophys. Res. D90:13076
39. Rinsland CP, Levine JS, Miles Th (1985) Nature 318:245
40. Rasmussen RA, Khalil MAK (1984) J. Geophys. Res. D89:11599
41. Craig H, Chou CC (1982) Geophys. Res. Lett. 9:1221
42. Pearman GI, Etheridge D, De Silva F, Fraser PJ (1986) Nature 320:248
43. Blake DR, Woo VH, Tyler SC, Rowland FS (1984) Geophys. Res. Lett. 11:1211
44. Altwicker ER, Whitby RA, Lioy PJ (1980) J. Geophys. Res. C85:7475
45. Rudolph J, Ehhalt DH (1981) J. Geophys. Res. C86:11959
46. Singh HB, Salas LJ (1982) Geophys. Res. Lett. 9:842
47. Cronn D, Robinson E (1979) Geophys. Res. Lett. 6:641
48. Goldman A, Murcray FJ, Blatherwick RD, Gillis JR, Bonomo FS, Murcray FH, Murcray DG, Cicerone RJ (1981) J. Geophys. Res. C86:12143
49. Rudolph J, Ehhalt DH, Khedin A (1984) J. Atmos. Chem. 2:117
50. Ehhalt DH, Rudolph J, Meixner F, Schmidt U (1985) J. Atmos. Chem. 3:29
51. Saiki Y, Suyama Y, Kashimura H, Wakamatsu S (1985) J. Jap. Soc. Air Pollut. 20:179
52. Investigation of Rural Oxidant Levels as Related to Urban Hydrocarbon Control Strategies (1975) US Environmental Protection Agency, EPA-450/3-75-036
53. Mayrsohn H, Crabtree JH (1976) Atmos. Environ. 10:137
54. Nassar J, Goldbach J (1979) Int. J. Environ. Anal. Chem. 6:145
55. Bruckmann P, Beier R, Krautscheid S (1983) Staub-Reinhalt. Luft 43:404
56. Lahman E, Seifert B, Dulson W (1978) Bundesgesundheitsbl. 21:75
57. Colbeck J, Harrison RM (1985) Atmos. Environ. 19:1899
58. Nelson PF, Quigley SM (1982) Environ. Sci. Techn. 16:650
59. Netravalkar AJ, Mohan Rao AM (1984) Sci. Total Environ. 35:33
60. Isaksen ISA, Hov Ø, Penkett SA, Semb A (1985) J. Atmos. Chem. 3:3

61. Rasmussen RA, Khalil MAK, Fox RJ (1983) Geophys. Res. Lett. 10:144
62. Grinberg JP, Zimmerman PR (1984) J. Geophys; Res. D89:4767
63. Rudolph J, Ehhalt DH, Tönnissen A (1981) J. Geophys. Res. C86:7267
64. Goldman A, Rinsland CP, Murcray FJ, Murcray DG, Coffey MT, Mankin WG (1984) J. Atmos. Chem. 2:221
65. Coffey MT, Mankin WG, Goldman A, Rinsland CP, Harvey GA, Devi VM, Stones JM (1985) Geophys. Res. Lett. 12:199
66. Blake DR, Rowland FS (1986) Nature 321:231
67. Eichman R, Ketseridis G, Schebeske G, Jaenicke R, Hahn J, Warneck P, Junge C (1980) Atmos. Environ. 14:695
68. Lonnenman WA, Seila RL Bufalini JJ (1978) Environ. Sci. Technol. 12:459
69. Churkin SP, Barakov TV, Stepen 'RA, Chernyaeva GN (1976) Khim. Prirodn. Soedin. N.2:260
70. Stepen 'RA, Konev VA, Khrebtov BA (1978) Izvestiya Sib. Otdel. AN SSSR. Ser. biol. nauk N 10:34
71. Isidorov VA, Zenkevich IG, Ioffe BV (1985) Atmos. Environ. 19:1
72. Rasmussen RA (1972) J. Air. Pollut. Contr. Assoc. 22:537
73. Whitby RA, Coffey PE (1977) J. Geophys. Res. 82:5928
74. Yokouchi Y, Fujii T, Ambe Y, Fuwa K (1981) J. Chromatogr. 209:293
75. Hov Ø, Schjoldager J, Wathne B (1983) J. Geophys. Res. C88:10679
76. Hutte RS, Williams EJ, Staehelin J, Hawthorne SB, Barkley RM, Sievers RE (1984) J. Chromatogr. 302:173
77. Ciccioli P, Brancaleoni E, Possanzini M, Brachetti A, Palo Di C (1984) Sci. Total. Environ. 36:255
78. Jüttner F (1986) Chemosphere 15:985
79. Burkser ES, Daen MI, Yuzefovich EK (1940) Vopr. kurortologii 4:36
80. Holdern MW, Westberg HH, Zimmerman PR (1979) J. Geophys. Res. 84:5082
81. Show RW. Crittenden AL, Stevens RK, Cronn DR, Titov VS (1983) Environ. Sci. Technol. 17:389
82. Riba ML. Tathy JP, Tsiropoulos N, Monsarrat B, Torres L (1987) Atmos. Environ. 21:191
83. Robinson E, Rasmussen RA, Westberg HH, Holdren MW (1973) J. Geophys. Res. 78:5345
84. Thorburn S, Colenutt BA (1979) Int. J. Environ. Stud. 13:265
85. Grob K, Grob G (1971) J. Chromatogr. 62:1
86. Singh HB, Salas LJ, Smith AJ, Shigeishi H (1981) J. Atmos. Environ. 15:601
87. Bertsch W, Chang RC, Zlatkis A (1974) J. Chromatogr. Sci. 12:175
88. Raymond A, Guiochon G (1974) Environ. Sci. Technol. 8:143
89. Ciccioli P, Bertoni G. Brancaleoni E, Fratarcangeli R, Bruner F (1976) J. Chromatogr. 126:757
90. Holzer G, Shanfield H, Zlatkis A, Bertsch W, Juarez P, Mayfield H, Liebich HM (1977) J. Chromatogr. 142:755
91. Ioffe BV, Isidorov VA, Zenkevich IG (1977) J. Chromatogr. 142:787
92. Ioffe BV, Isidorov VA, Zenkevich IG (1979) Environ. Sci. Technol. 13:864
93. Low CW, Richards JF, Faure PK (1977) Atmos. Environ. 11:703
94. Hagemann R, Virelizer H, Gaudin D (1978) Analysis 6:401
95. Dmitriev MT, Rastyannikov EG, Gladkov BC (1982) Trudi Zentr. Visotn. Gidrometeorol. Observ. N 16:31
96. Isidorov VA, Zenkevich IG, Ioffe BV (1983) Atmos. Environ. 17:1347

97. Ioffe BV, Isidorov VA, Zenkevich IG (1978) Dokl. AN SSSR 243:1186
98. Isidorov VA, Zenkevich IG, Ioffe BV (1981) Gigiena i Sanit. N 1:19
99. Isidorov VA, Perez RG, Ioffe BV (1986) Vestnik Leningr. Univers. Ser. 4N 2:64
100. Jeltes R, Burghardt E, Thijsse TR (1977) Chromatographia 10:430
101. Isidorov VA, Zenkevich IG (1982) Mass Spectrometric-Gas Chromtographic Determination of Traces of Organic Components in Atmosphere. Khimiya, Leningrad.
102a. Dhar NR, Ram A (1932) Nature 130:313
102b. Lowe DC, Schmidt U, Ehhalt DH (1980) Geophys. Res. Lett. 7:825
103. Platt U, Perner D (1980) J. Geophys. Res. C85:7453
104. Lowe DC, Schmidt U, Ehhalt DH, Frischkorn CGB, Nürnberg HW (1981) Environ. Sci. Technol. 15:819
105. Lowe DC, Schmidt U (1983) J. Geophys. Res. C88:10844
106. Fushimi K, Miyake Y (1980) J. Geophys. Res. C85:7533
107. Zafiriou OC, Alford J, Herrera M, Peltzer ET, Gagosian PB (1980) Geophys. Res. Lett. 7:341
108. Schulam P, Newbold R, Hull LA (1985) Atmos. Environ. 19:623
109. Tanner RL, Meng Z (1984) Environ. Sci. Technol. 18:723
110. Grosjean D, Swanson RD, Ellis C (1983) Sci. Total Environ. 29:65
111. Grosjean D, Fung K (1984) J. Air Pollut. Contr. Assoc. 34:537
112. Grosjean D (1982) Environ. Sci. Technol. 16:254
113. Salas LJ, Singh HB (1986) Atmos. Environ. 20:1301
114. Cleviland WS, Graedel TE, Kleiner B (1977) Atmos. Environ. 11:357
115. Kuwata K, Uebori M, Jamasaki Y (1979) J. Chromatogr. Sci. 17:264
116. Van Neste A, Duce RA (1987) Geophys. Res. Lett. 14:711
117. Becker KH, Ionescu A (1982) Geophys. Res. Lett. 9:1349
118. Snider JR, Dawson GA (1984) Geophys. Res. Lett. 11:241
119. Brasseur G, Zellner R, De Rudder A, Arijs E (1985) Geophys. Res. Lett. 12:117
120. Arnold F, Hauck G (1985) Nature 315:307
121. Snider JB, Dawson GA (1985) J. Geophys. Res. D90:3797
122. Dawson GA, Farmer JC, Moyers JL (1980) Geophys. Res. Lett. 7:725
123. Kawamura K, Kaplan JR (1984) Anal. Chem. 56:1616
124. Goldman A, Murcray FH, Murcray DG, Rinsland CP (1984) Geophys. Res. Lett. 11:307
125. Norton RB (1985) Geophys. Res. Lett. 12:769
126. Keene WC, Galloway JN (1986) J. Geophys. Res. D91:466
127. Andreae MO, Talbot RW, Li Shao-Meng (1987) J. Geophys. Res. D92:6635
128. Grosjean D (1983) Environ. Sci. Technol. 17:13
129. Andreae MO, Andreae TW (1988) J. Geophys. Res. D93:1487
130. Andreae MO, Ferek RJ, Bermond F, Byrd KP, Engstrom RT, Hardid S, Houmere PD, Le Marret F, Raemdonck H, Chatfield RB (1985) J. Geophys. Res. D90:12891
131. Ferek RJ, Chatfield RB, Andreae MO (1986) Nature 320:514
132. Fine DH, Fan S, La Fleur A (1980) AIChE Symp. Ser. 76:305
133. Williams JH (1965) Anal. Chem. 37:1723
134. Saunders RA, Griffith JR, Saalfield FE (1974) Biomed. Mass. Spectrom. 1:192
135. Bunn WW, Deane ER, Klein DW, (1975) Water, Air Soil Pollut. 4:367
136. Jonson L, Josefsson B, Marstorp P (1981) Int. J. Environ. Anal. Chem. 26:7
137. Arnold F, Knop G, Ziereis H (1986) Nature 321:505

138. Chuong BT, Bodin D, Benarie M (1978) Assessment of Dimethyl-N-nitrosamine in urban atmosphere. In: Atmos. Pollut. 1978. Amsterdam pp 73–76
139. Löbel J, Wipprecht V, Schurath U (1980) Staub-Reinhalt. Luft. 40:243
140. Singh HB, Salas LJ, Ridley BA, Shetter JD, Donahue NM, Febsenfeld FS, Fahey DW, Parrish DD, Williams EJ (1985) Nature 318:347
141. Singh HB, Salas LJ, Viezee W (1986) Nature 321:588
142. Bottenheim JW, Gallant AG, Brice RA (1986) Geophys. Res. Lett. 13:113
143. Ingels J, Nevejans D, Frederick P, Arijs E (1986) Aeron. Acta N 311:1
144. Murad E, Swider W, Moss RA, Toby S (1984) Geophys. Res. Lett. 11:147
145. Bingemer H (1982) Measurements of Reduced Sulfur Gases in the Atmosphere. In: Atmos. Trace Constituents. Proc. 5 Two-Annu. Colloq. Mainz. 1981. Braunschweig Wiesbaden pp 3–15
146. Andreae MO, Raemdonck H (1983) Science 221:744
147. Molina MJ, Rowland FS (1974) Nature 249:810
148. Rowland FS, Molina MJ (1975) Rev. Geophys. Space Phys. 13:1
149. Grimsrud EP, Rasmussen RA (1975) Atmos. Environ. 9:1014
150. Rasmussen RA, Rasmussen LE, Khalil MAK, Dalluge RW (1980) J. Geophys. Res. C85:7350
151. Rasmussen RA, Khalil MAK, Penkett SA, Prosser NJD (1980) Geophys. Res. Lett. 7:809
152. Singh HB, Salas LJ, Stiles RE (1982) Environ. Sci. Technol. 16:872
153. Penkett SA, Prosser NJD, Rasmussen RA, Khalil MAK (1981) J. Geophys. Res. C86:5172
154. Singh HB, Salas LJ, Stiles RE (1983) J. Geophys. Res. C88:3675
155. Rasmussen RA, Khalil MAK (1983) Antaratic J. US 18:250
156. Fabin P, Borchers R, Krüger BC, Lal S, Penkett SA (1985) Geophys. Res. Lett. 12:1
157. Khalil MAK, Rasmussen RA (1985) Geophys. Res. Lett. 12:671
158. Class Th, Ballschmiter K (1987) Fresenius Z. Anal. Chem. 327:198
159. Cronn DR, Bamesberger WL, Menzia FA, Waylett SF, Ferrari TW, Howard HM, Robinson E (1986) Geophys. Res. Lett. 13:1272
160. Cunnold DM, Prinn RG, Rasmussen RA, Simmonds PG, Alyea FN, Cardelino CA, Crawford AJ, Fraser PJ, Rosen RD (1986) J. Geophys. Res. D91:10797
161. Yokohata A, Makide Y, Tominaga T (1984) Mem. Nat. Inst. Polar Res. Spec. N 34:231
162. Rasmussen RA, Khalil MAK (1984) Geophys. Res. Lett. 11:433
163. Berg WW, Heidt LE, Pollack W, Sperry PD, Cicerone RJ (1984) Geophys. Res. Lett. 11:429
164. Penkett SA, Jones BMR, Rycroft MJ, Simmons DA (1985) Nature 318:550
165. Cronn DR, Rasmussen RA, Robinson E, Harsch DE (1977) J. Geophys. 82:5935
166. Makide Y, Tominaga T, Rowland FS (1979) Chem. Lett. 4:355
167. Singh HB, Martinez JR, Hendry DG, Jaffe RT (1981) Environ. Sci. Technol. 15:113
168. Singh HB, Salas LJ, Cavanagh LA (1977) J. Air Pollut. Contr. Assoc. 27:332
169. Mann JB, Freal JJ, Enos HF, Danauskas JX (1980) J. Environ. Sci. Health B15:507
170. Robinson E, Bamesberger WL, Menzia FA, Waylett AS, Waylett SF (1984) J. Atmos. Chem. 2:65
171. Pack DH, Lovelock JE, Cotton G, Curthoys C (1977) Atmos. Environ. 11:329
172. Prinn RG, Simmonds PG, Rasmussen RA, Rosen RD, Alyea FN, Cardelino CA,

Crawford AJ, Cunnold DM, Fraser PJ, Lovelock JE (1983) J. Geophys. Res. C88:8353
173. Khalil MAK, Rasmussen RA (1984) Chemosphere 13:789
174. Schmidt U, Khedim A, Knapska D, Kulessa G, Johnes FJ (1984) Adv. Space. Res. 4:131
175. Schmidt U, Knapska D, Penkett SA (1985) J. Atmos. Chem. 3:363
176. Gallagher CC, Forsberg CA, Mason AS, Gandrud BW, Langhorbani M (1985) J. Geophys. Res. D90:10747
177. Vedder JF, Tyson BJ, Brewer RB, Boitnott CA, Inn ECY (1978) Geophys. Res. Lett. 5:33
178. Afanasiev MI, Vulykh NK, Zagruzina AN, Teplizkaya TA, Alekseeva TA (1986) Background Content of Organochlorine Pesticides and Polycyclic Aromatic Hydrocarbons in Natural Environments (According to World Data). Communication 3. In: Monitoring of Background Pollution of Natural Environments. Gidrometeoizdat, Leningrad pp. 27–53
179. Bidleman TE, Christensen EJ, Billings WN, Leonard R (1981) J. Marin. Res. 39:443
180. Yasuda H (1980) Chem. Educ. (Jap) 28:75
181. Tanabe S, Hidaka H, Tatsukawa R (1983) Chemosphere 12:277
182. Kawano M, Tanabe S, Inoue T, Tatsukawa R (1985) Trans. Tokyo Univ. Fish. N 6:59
183. Rohbock E, Georg HW, Müller J (1980) Atmos. Environ. 14:89
184. De Jongle WRA, Chakroborti D, Adams FS (1981) Environ. Sci. Techn. 15:1217
185. Harrison RM, Perry R, Slater DH (1974) Atmos. Environ. 8:1187
186. Harrison RM, Hewitt CN, Radojevich M (1985) Environmental Parthways of Alkyllead Compounds. In: Heavy Metals Environ. Int. Conf. Athens. Sept. 1985 Vol. 1. Edinburgh, pp. 82–84
187. Hewitt CN, Harrison RM, Radojevic M (1986) Anal. Chem. Acta 188:247
188. Lindqvist O, Rodhe H (1985) Tellus B37:136
189. Paudyn A, Van Loon JC (1986) Fresenius Z. Anal. Chem. 325:369
190. Howard AG, Arbab-Zavar MH, Apte S (1982) Marin. Chem. 11:493
191. Cooke TD, Bruland KW (1987) Environ. Sci. Technol. 21:1214
192a. Jiang S, Robberecht H, Adams F (1983) Atmos. Environ. 17:111
192b. Cadle RD (1973) Suspected particles in the lower atmosphere. In: Rasool SI (ed) Chemistry of the lower atmosphere. Plenum Press, New York London
192c. Goetz A, Pueschel R (1967) Atmos. Environ. 1:287 (instead Goetz and Preining)
193. Duce RA, Mohnen VA, Zimmerman PR, Grosjean D, Cautreels W, Chatfield R, Jaenicke R, Ogren JA, Pellizzani ED, Wallace GT (1983) Rev. Geophys. Space Phys. 21:921
194. Broddin G, Cautreels W, Van Cauwenberghe K (1980) Atmos. Environ. 14:895
195. Neuling P, Neeb R, Eichman R, Junge C (1980) Z. Anal. Chem. 302:375
196. Grosjean D, Fung K, Mueller P, Heisler S, Hidy G (1980) AIChE Symp. Ser. 76:96
197. Hidy GM (1986) Definition and Characterization of Suspended Particles in Ambient Air. In: Aerosols: Res., Risk Assess. and Contr. Strateg. Proc. 2nd US-Dutsh Int. Symp. Williamsburgh, Va. May 19–25, 1985. Chelsea, Mich. pp. 19–41
198. Katz M, Chan C (1980) Environ. Sci. Technol. 14:838
199. Simoneit BRT, Mazurek MA (1982) Atmos. Environ. 16:2139
200. Simoneit BRT (1984) Atmos. Environ. 18:61
201. Simoneit BRT (1984) Sci. Total Environ. 36:61

202. Lioy PJ, Kneip ThJ, Daisey JM (1984) J. Geophys. Res. D 89:1355
203. Stevens RK, Dzubay TG, Shaw RW, McCenny WA, Lewis CW, Wilson WE (1980) Environ. Sci. Technol. 14:1491
204. Ketseridis G, Hahn J, Jaenike R, Junge C (1976) Atmos. Environ. 10:603
205. Grosjean D (1984) Sci. Total Environ. 32:133
206. Mueller PR, Mosley RW, Pierce LB (1982) J. Coll. Interface Sci. 39:235
207. Bravo JI, Salazar S (1984) Geofis. int. 24:476
208. Lee RE, Jr, Hein J, (1974) Anal. Chem. 46:931
209. Simoneit BRT (1977) Marin. Chem. 5:433
210. Halkiewicz J, Lamparczyk H, Grzybowski J, Radecki A (1987) Atmos. Environ. 21:2057
211. Sicre MA, Marty JC, Saliot A, Aparicio X, Grimalt J, Albaiges J (1987) 21:2247
212. Van Vaeck L, Broddin G, Van Cauwenberghe K (1979) Environ. Sci. Technol. 13:1494
213. Boone PM, Macias ES (1987) Environ. Sci. Technol. 21:903
214. Schuetzle D, Lee FSC, Prater TJ, Tejada SB (1981) Int. J. Environ. Anal. Chem. 9:93
215. König J, Funcke W, Balfanz E, Grosch R, Romanowski Th (1981) Staub-Reinhalt Luft. 41:79
216. Marty JC, Tissier MJ, Saliot A (1984) Atmos. Environ. 18:2183
217. Arey J, Zielinska B, Atkinson R, Winer AM (1987) Atmos. Environ. 21:1437
218. Daisey JM, McCaffey RJ, Gallagher RA (1981) Atmos. Environ. 15:1553
219. Teplizkaja TA (1986) Background monitoring of polycyclic aromatic hydrocarbons. In: Complex Global Monitoring of Biosphere. Proc. 3nd Int. Symp. Tashkent, 14–19 Oct 1985. Vol 1 Gidrometeoizdat Leningrad, pp 144–154
220. Colmsjö AL, Zebühr YU, Ostman CE, Wädding A, Söderström H (1986) Chemosphere 15:169
221. Guicherit R, Schulting FL (1985) Sci. Total Environ. 43:193
222. Jaklin J, Krenmayer P (1985) Int. J. Environ. Anal. Chem. 21:33
223. Muel C, Saguem S (1985) Int. J. Environ. Anal. Chem. 19:111
224. Kawamura K, Gagosian RB (1987) J. Chromatogr. 390:371
225. Van Vaeck, Van Cauwenberghe K (1978) Atmos. Environ. 12:2229
226. Karasek FW, Denney DW, Chan KW, Clement RE (1978) Anal. Chem. 50:82
227. Gill PS, Graedel TE, Weschler CJ (1983) Rev. Geophys. Space Phys. 21:903
228. Efimova AA (1980) Productivity of natural continental plants as an element of carbon dioxide cycle. In: The Problems of Atmosphere Carbon Dioxide. Proc. Soviet-Amer. Symp. Dushanbe, 12–20 Oct 1978 Gidrometeoizdat. Leningrad, pp 79–85
229. Mirchink TG, Pannikov NS (1985) Uspechi mikrobiol. 20:198
230. Pankratova EM (1987) Uspechi mikrobiol. 21:212
231. Nilov VI (1928) Nauchno-agron. Zh. (J. Landwirtschaft. Wissenshaft) No 1:810
232. Bryantzeva ZI (1951) Uchen. Zap. Gorkov. Univers. 21:149
233a. Artemjeva MN (1962) Vopr. Klimatol. 4:289
233b. Rice EL (1974) Allelopathy. Academic Press, New York San Francisco London
233c. Sanadze GA (1960) Emission of volatile organic compounds from vegetations. Izdatelstvo AN Gruz. SSR, Tbilisi
234. Isidorov VA, Zenkevich IG, Ioffe BV 9 (1982) Dokl. AN SSSR 263:893
235. Khasanov AA, Isidorov VA, Zenkevich IG, Ioffe BV (1982) Dokl. AN UzSSR No 12:29

236. Went F (1966) Tellus 18:549
237. Zimmerman PR, Chatfield RB, Fishman J, Crutzen PJ, Hanst PL (1978) Geophys. Res. Lett. 5:679
238. Peterson EW, Tingey D (1980) Atmos. Environ. 14:79
239. Flyckt DL, Westberg HH, Holdren MW (1980) Natural organic emissions and their impact on air quality. In: Proc. 73rd Annu. Meet. Montreal, June 22–27 Vol. 5. Pittsburgh, Pa. 69. 2/1–69.2/14
240. Knoppel H, Versino B, Pell A, Schauenburgh H, Uissers H (1982) Quantitative determination of terpenes emitted by conifers. In: Phys–Chem. Behav. Atmos. Pollutants. Proc. 2nd Eur. Symp. Varese, 29 Sept 1981. Dordrecht pp 89–98
241. Cronn D, Nutmagul W (1982) Tellus 34:159
242a. Protopopov VV (1975) The significance of dark coniferous forest in the formation of its environment. Nauka, Novosibirsk
242b. Persson R (1974) World forest resources. Dep. of forest survey. Royal College of forestry, Stockholm
242c. Went FW (1960) Proc. Nat. Acad. Sci. 46:212
242d. Rasmussen RA, Went FW (1965) Proc. Nat. Acad. Sci. 53:215
243. Lamb B, Wesberg H, Allwine G, Quarles T (1985) J. Geophys. Res. D90:2380
244. Whittaker RH, Likens G (1975) The biosphere and man. In: Lieth H, Whittaker RH (eds) Primary Productivity of the Biosphere. Springer, Berlin Heidelberg New York pp. 305–328
245. Productivity of organic and biologic forest mass (1974) Molchanov AA (ed) Nauka, Moscow
246. Lamb B, Guenther A, Gray D, Westberg H (1987) Atmos. Environ. 21:1695
247. Dimitriades B (1981) J. Air Pollut. Contr. Assoc. 31:229
248. Lamb B, Westberg H, Allwine G (1986) Atmos. Environ. 20:1
249. Gshwend P, Zafiriou OC, Gagosian RB (1980) Limnol. Oceanogr. 25:1044
250. Sauer TC, Jr (1981) Org. Geochem. 3:91
251a. Rasmussen RA, Khalil MAK, Gunawardena R, Hoyt SD (1982) J. Geophys. Res. C87:3086
251b. Lovelock JE, Maggs RJ, Rasmussen RA (1972) Nature 237:452
252. Class T, Ballschmiter K (1987) Fresenius Z. Anal. Chem. 327:40
253. Andreae MO (1985) The emission of sulfur to remote atmosphere. In: Biogeochem. Cycling Sulfur and Nitrogen in the Remote Atmosphere. Dordrecht. pp 3–25
254. Toon OB, Kasting JF, Turco RP, Liu MS (1987) J. Geophys. Res. D92:943
255. Cline JD, Bates TS (1983) Geophys. Res. Lett. 10:949
256. Andreae MO, Raemdonck H (1983) Science 221:744
257. Stotzky G, Schenck S (1976) C.R.C. Crit. Revs. Microbiol. 4:338
258. Sheppard JC, Westberg H, Hopper JF, Ganesan K, Zimmerman P (1982) J. Geophys. Res. C87:1305
259. Seiler W (1982) The cycle of CH_4 in the troposphere. In: AMS-AGU-NASA Conf. Williamsburg, Va. May 25–26
260. Cicerone RJ, Walters S, Liu SC (1983) J. Geophys. Res. C88:3647
261. Seiler W, Conrad R, Scharffe D (1984) J. Atmos. Chem. 1:171
262. Keller M, Kaplan WA, Wofsy SC (1986) J. Geophys. Res. D91:11791
263. Bartlett KB, Harriss RC, Sebacher DI (1985) J. Geophys. Res. D90:5710
264. Harriss RC, Gorham E, Sebacher DI, Bartlett KB, Flebbe PA (1985) Nature 315:652
265. Sebacher DI, Harriss RC, Bartlett KB, Sebacher SM, Grice SS (1986) Tellus B38:1

266. Holzapfel-Pschorn A, Seiler W (1986) J. Geophys. Res. D91:11804
267. Fraser PJ, Rasmussen RA, Creffield JW, French JR, Khalil MAK (1986) J. Atmos. Chem. 4:295
268. Min'ko IO (1988) Pochvovedenie No 7:59
269. Orlov DS, Min'ko OI, Kasparov SV, Ammosova YaM (1987) Pochvovedenie No 6:89
270. Adams DF, Farwell SO, Robinson E, Pack MR, Bamesberger WL (1981) Environ. Sci. Technol. 15:1493
271. Aneja VP, Overton JH, Aneja AP (1981) J. Air Pollut. Contr. Assoc. 31:256
272. Steudler PA, Peterson BJ (1985) Atmos. Environ. 19:1411
273. Pereira WE, Rostand CE, Taylor HE (1980) Geophys. Res. Lett. 7:953
274. Isidorov VA, Zenkevich IG (1985) Dokl. AN SSSR 280:223
275. Isidorov VA, Titov VS, Strukova TP (1985) Meteorol. i Gidrol. No 10:229
276. Isidorov VA, Ioffe BV (1986) Dokl. AN SSSR 287:86
277a. Bondarev VB, Porshnev NV (1980) Dokl. AN SSSR 252:455
277b. Beskrony NS, Lobkov VA (1977) Relationships of spreading of hydrocarbon gases in modern hydrothermal systems of Kamchatka. In: Geothermal process in the regions of tectonic-magmatic activity. Nauka; Moscow
278. Menyailov IA, Nikitina LP, Shapar' VN (1986) Vulkanol. i Seismol. No 4:43
279. Khalil MAK, Rasmussen RA (1982) Chemosphere 11:877
280. Rasmussen RA, Khalil MAK, Dalluge RW, Penkett SA, Jones B (1982) Science 215:665
281. Zor'kin LM, Korzenstein YN, Stadnik EV, Kozlov VG, Kir'yashin VM, Yurin GA, Borodkin VA (1980) Dokl. AN SSSR 252:681
282. Voitov GI, Mil'kes MR, Kucher MI, Belikov VM, Krivomazova NG (1986) Dokl. AN SSSR 290:1335
283. Fabian P, Borshers R, Krüger BC, Lal S (1987) J. Geophys. Res. D92:9831
284. Crutzen PJ, Heidt LE, Krasnec JP, Pollock WH, Seiler W (1979) Nature 282:253
285. Westberg H, Sexton K, Flyckt D (1981) J. Air Pollut. Contr. Assoc. 31:661
286. Greenberg JP, Zimmerman PR, Heidt L, Pollock W (1984) J. Geophys. Res. D89:1350
287. Edvards PR, Campbell I, Milne GS (1982) Chem. Ind. No 16:574
288. Zinger DE (1985) SAE Techn. Pap. Ser. No 851262
289. Murakami Y, Nishida K, Yamakawa M (1985) J. Environ. Pollut. Contr. 21:635
290. Kashmiura H, Suyama Y, Saiki Y, Yamamoto A, Himi Y (1983) J. Jap. Soc. Air Pollut. 18:432
291a. Hampton CV, Pierson WR, Harvey TM, Updegrove WS, Morana RS (1982) Environ. Sci. Technol. 16:287
291b. Seizinger DE, Dimitriades B (1977) J. Air Pollut. Contr. Assoc. 22:47
292. Matthews RD (1980) J. Combust. Toxicol. 7:157
293. Urban CM, Garbe RJ (1979) SAE Techn. Pap. Ser. No 790696
294. Ballschmiter K, Mayer P (1983) Chemosphere 12:879
295. Yu Ming-Li, Hites RA (1981) Anal. Chem. 53:951
296a. Morita K, Fukamachi K, Tokiwa H (1982) Bunseki kagaky 31:255
296b. Words K (1980) J. Environ. Sci. Health A15:573
297. Khesina A. Ya, Smirnov GA, Shabad LM, Prich B, Nauman M (1981) Contribution of automobile transport to the pollution of urban air by PAH. In: Snizhenie toksichnosti otrabotavshikh gasov dvigatelei vnutrennego sgoraniya. Moscow, pp 37–42

298a. Benkovitz CM (1982) Atmos. Environ. 16:1551
298b. Lee K-C, Hansen JL, Whipple GMA (1980) A fugitive emission study in a petrochemical manufacturing unit. In:Proc. 73 Annu. Meet. Montreal. June 22–27, V. 5. Pittsburgh, Pa, s.a. 689/1–689/16
299a. Häsänen E, Pohjola V, Hahkala M, Zilliacus R, Wickström K (1986) Sci. Total Environ. 54:29
299b. Manual on the hygiene of atmospheric air (1976) Buschtueva KA (ed). Meditsina, Moscow
300. Knapp KT, Bennett RL, Jones PW (1980) AIChE Symp. Ser. 76:323
301. Sartorius R, Jost D (1979) Umshau Wiss. Techn. 79:193
302. Leifer R, Larsen R, Toonkee L (1982) Geophys. Res. Lett. 9:755
303. Götz H (1979) Gas, Wasser, Wärme 33:195
304. Höfler F, Schneider J, Möckel HJ (1986) Fresenius Z. Anal. Chem. 325:365
305. Angrick M (1987) Müll and Abfall 19:142
306a. Bingemer HG, Crutzen PJ (1987) J. Geophys. Res. D92:2181
306b. Vanni A, Esposito A (1982) Landfill gas recovery:state of the art in Italy; a feasibility study for one of the biggest Italian landfills. In:Resour. Recov. Solid Wastes. Proc. Conf. Miami Beah, FL, May 10–12, 1982 New York, pp 417–424
307. Wilkins ES, Wilkins MG (1985) J. Environ. Sci. Health A20:149
308. Hawley-Fedder RA, Parsons ML, Karasek FW (1987) J. Chromatogr. 387:207
309. Hutzinger O, Blumich MJ, Berg M, Olie K (1985) Chemosphere 14:581
310. Sheffield A (1985) Chemosphere 14:811
311. Marklund S, Rappe Ch, Tysklind M, Egebäck K-E (1987) Chemosphere 16:29
312. Eiceman GA, Clement RE, Karasek FW (1981) Anal. Chem. 53:955
313a. Eastman JA, Stedman DH (1980) Atmos. Environ. 14:731
313b. Hewitt CN, Harrison RM (1985) Atmos. Environ. 19:545
313c. McIven M, Phillips L. Chemistry of the Atmosphere, 1975
314. Lewin EE, De Pona RG, Shimshock JP (1986) Atmos. Environ. 20:59
315. Harris GW, Carter WPL, Winer AM, Pitts JN, Jr, Platt U, Perner D (1982) Environ. Sci. Technol. 16:414
316. Atkinson R, Aschmann SM, Winer AM (1987) J. Atmos. Chem. 5:91
317. Baulch DL, Cox RA, Hampton RF, Jr, Kerr JA, Troe J, Watson RT (1984) J. Phys. Chem. Ref. Data 13:1259
318. Atkinson R, Llyod AC (1984) J. Phys. Chem. Ref. Data 13:315
319. Akimoto H, Sakamaki F (1983) Environ. Sci. Technol. 17:94
320. Hanst PL, Spence JW, Edney EO (1980) Atmos. Environ. 14:1077
321. Schuetzle D, Rasmussen RA (1978) J. Air. Pollut. Contr. Assoc. 28:236
322. Killus JP, Whitten GZ (1984) Environ. Sci. Technol. 18:124
323. Atkinson R, Aschmann SM, Winer AM, Pitts JN, Jr, (1984) Environ. Sci. Technol. 18:370
324. Atkinson R, Aschmann SM, Winer AM, Pitts JN, Jr (1985) Environ. Sci. Technol. 19:159
325. Gu Chee-Ilang, Rynard CM, Hendry DJ, Mill T (1985) Environ. Sci. Technol. 19:151
326. Cachier H, Buat-Menard P, Fontugne M, Rancher J (1985) J. Atmos. Chem 3:469
327. Lopez A, Fontan J, Bartomeuf MO (1985) J. Rech. Atmos. 19:295
328. Golovina EG, Ivlev LS, Sirota VG (1985) Meteorol. i gidrol. N 11:107
329. Shulanov YuV (1986) Izvestiya AN SSSR. FAO 22:616

330. Dumdei B, O'Brien RJ (1984) Nature 311:248
331. Leone JA, Flagan RC, Grosjean D, Seinfeld JH (1985) Int. J. Chem. Kinet. 17:177
332. Tuazon EC, McLeod H, Atkinson R, Carter WPL (1986) Environ. Sci. Technol. 20:383
333. Nojima K, Kawaguchi A, Ohaya T, Kanno S, Hirobe M (1983) Chem. and Pharm. Bull. 31:1047
334. Stern JE, Flagan RC, Grosjean D, Seinfeld JH (1987) Environ. Sci. Technol. 21:1224
335. Horowitz A (1985) J. Phys. Chem. 89:1764
336. Gardner EP, Wijayaratne RD, Calvert JG (1984) J. Phys. Chem. 88:5069
337. Zetzsch C, Stuhl F (1982) Rate constants for reactions of OH with carbonic acids. In: Phys. Chem. Behav. Atmos. Pollutants. Proc. 2nd Eur. Symp. Varese. Dordrecht pp 129–137
338. Harris GW, Pitts JN, Jr (1983) Environ. Sci. Technol. 17:50
339. Koda S, Yoshikawa K, Okada J, Akita K (1985) Environ. Sci. Technol. 19:262
340. Yin Fangdong, Grosjean D, Seinfeld JH (1986) J. Geophys. Res. D91:417
341. Saltzman ES, Savoie DL, Prospero JM, Zika RG (1986) J. Atmos. Chem. 4:227
342. Ayers GP, Jvey JP, Goodman HS (1986) J. Atmos. Chem. 4:173
343a. Atkinson R, Pitts JN, Aschmann SM (1984) J. Phys. Chem. 89:1584
343b. Haagen-Smith AJ (1952) Ind. Eng. Chem. 44:1342
344. Atkinson R, Lloyd A, Winges L (1982) Atmos. Environ. 16:1341
345. Gery MW, Fox DL, Jeffries HE, Stockburger L, Weathers WS (1985) Int. J. Chem. Kinet. 17:931
346a. Gery MW, Fox DL, Kamens RM, Stockburger L (1987) Environ. Sci. Technol 21:339
346b. Fox DL, Gerry MW, Jeffries HE (1980) Organic aerosol formation mechanisms–a smog chamber study. In: Proc 73 Annu. Meet. Air Pollut. Contr. Assoc. Montreal June 22–27, 1980, v. 4 Pittsburgh, Pa, 80–50.2/1–80–50.2/16
347a. McMurry PH, Fridlander SK (1979) Atmos. Environ. 13:1635
347b. Crutzen PJ (1970) Quant. J. Roy. Meteorol. Soc. 96:320
347c. Johnston HS (1971) Science 173:517
348. Spence JW, Hanst PL (1978) J. Air Pollut. Contr. Assoc. 28:250
349a. Yung YL, Pinto JP, Watson RT, Sander SP (1980) J. Atmos. Sci. 37:339
349b. Wofsy SC, McElroy MB, Yung YL (1975) Geophys. Res. Lett. 2:215
350. Chuan RL, Woods DC (1984) Geofis. int. 23:335
351. Zemtzov AN (1986) Investigation of the solid disperse phase of the volcanic eruptive cloud. Kand. Diss. Moscow. Inst. of Lithosphere AN SSSR
352. Weisel CP, Duce RA, Fasching JL, Heaton RW (1984) J. Geophys. Res. D89:11607
353a. Anikiev VV, Il'ichev VL, Lobanov AA, Medvedev AN (1985) Dokl. AN SSSR 281:937
353b. Junge CE (1963) Air Chemistry and Radioactivity. Academic Press, New York
354. Fox MN (1983) Account. Chem. Res. 16:314
355. Takeuchi K, Yazawa T, Ibusuki T (1983) Atmos. Environ. 17:2253
356. Takeuchi K, Ibusuki T (1986) Atmos. Environ. 20:1155
357. Gäb S, Schmitzer J, Thamm HW, Parlar H, Korte F (1977) Nature 270:331
358. Ausloos P, Rebbert RE, Glasgow L (1977) J. Res. NBS 82:1
359. Kotzias D, Klein W, Lotz F, Nitz S, Korte F (1979) Chemosphere 8:301

360. Sancier KM, Wise H (1981) Atmos. Environ. 15:639
361. Anpo M, Nakaya H, Kodama S, Kubokawa Y, Domen K, Onishi T (1986) J. Phys. Chem. 90:1633
362. Tanaka T, Ooe M, Funabiki T, Yoshida S (1986) J. Chem. Soc. Faraday Trans. Pt. 1 82:35
363. Alekseev AV, Gerasimov SF, Pozdnyakov DV, Filimonov NV (1980) Uspechi Fotoniki 7:143
364. Pozdnyakov DV, Filimonov VN, Kondrate'v KYa (1980) Dokl. AN SSSR 252:1097
365. Filby WG, Mintas M, Güsten H (1981) Ber. Bunsenges. Phys. Chem. 18:189
366. Onuska FI, Karasek FW (1984) Open Tubular Column Gas Chromatography in Environmental Sciences. Plenum. New York
367. Bannach OS, Dazenko II, Timkovich AZ (1984) Gigiena i sanit. No. 2:53
368. Oomens AC, Noten LG (1984) J. High Resolut. Chromatogr. and Chromatogr. Commun. 7:280
369. Netravalkar AJ, Mohan Rao AM (1986) Chromatographia 22:183
370. Goldsmith (1986) LC and GC 4:939
371. Isidorov VA, Perez RG (1987) Vestnik Leningr. Univers. Ser. 4 No. 1:59
372. Ioffe BV, Vitenberg AG (1984) Head-Space Analysis and Related Methods in Gas Chromatography. John Wiley, New York
373. Kuznetzova LM, Isidorov VA (1989) Gigiena i sanit. 5:54
374. Isidorov VA, Kuznetzova LM, Maevskii GA (1984) Gas chromatographic determination of amines in atmospheric air. In:5 All-Union Conference Anal. Chem. Org. Compounds. Nauka, Moscow, p 117
375. Lindskog A, Brorström-Linden E, Alfheim I, Hagen I (1987) Sci. Total Environ. 61:51
376. Szepency L, Laksznerk K, Ackerman L, Podmaniczky L (1981) J. Chromatogr. 206:611
377. Crouch RL, Hawley-Pedder RA, Parsons ML, Karasek FW (1984) J. Chromatogr. 303:53
378. Liberti A, Ciccioli P (1986) J. High Resolut. Chromatogr. and Chromatogr. Commun. 9:492
379. Wright BW, Wright CW, Gale RW, Smith RD (1987) Anal. Chem. 59:38
380. Cooper JR, Taylor LT (1984) Anal. Chem. 56:1989
381. Røyset O, Thomassen Y (1986) Anal. Chim. Acta 188:247
382. Zander R; Stokes GM, Brault JW (1983) Geophys. Res. Lett. 10:521
383. Gabrielyan AG, Grechko EI, Dianov-Klockov VI (1983) Izvestiya AN SSSR. FAO 19:427
384. Garrigues P, Ewald M (1987) Chemosphere 16:485

Index

**The Handbook
of Environmental
Chemistry**

Editor: O. Hutzinger

Volume 1

*The Natural Environment
and the Biogeochemical Cycles*

Part A

1st ed. 1980. Corr. 2nd printing. XV, 258 pp. 54 figs.
Hardcover DM 148,– ISBN 3-540-09688-4

Part B

1982. XV, 317 pp. 84 figs. Hardcover DM 220,–
ISBN 3-540-11106-9

Part C

1984. XIII, 220 pp. 55 figs. Hardcover DM 140,–
ISBN 3-540-13226-0

Part D

1985. XI, 246 pp. 58 figs. Hardcover DM 192,–
ISBN 3-540-15000-5

Part E

1990. XI, 192 pp. 66 figs. 5 tabs. Hardcover DM 128,–
ISBN 3-540-15548-1

Volume 2

Reactions and Process

Part A

1980. XVIII, 307 pp. 66 figs. 27 tabs. Hardcover DM 148,–
ISBN 3-540-09689-2

Part B

1982. XV, 205 pp. 63 figs. Hardcover DM 128,–
ISBN 3-540-11107-7

Part C

1985. XIII, 145 pp. 49 figs. Hardcover DM 112,–
ISBN 3-540-13819-6

Part D

1988. XI, 210 pp. 47 figs. 55 tabs. Hardcover DM 186,–
ISBN 3-540-15547-3

Springer-Verlag Berlin
Heidelberg New York London
Paris Tokyo Hong Kong

Part E

1989. XI, 240 pp. 50 figs. 40 tabs. Hardcover DM 168,–
ISBN 3-540-51126-1

The Handbook of Environmental Chemistry

Editor: O. Hutzinger

Volume 3

Anthropogenic Compounds

Part A

1980. XV, 274 pp. 61 figs. Hardcover DM 116,–
ISBN 3-540-09690-6

Part B

1982. XVII, 210 pp. 38 figs. Hardcover DM 148,–
ISBN 3-540-11108-5

Part C

1984. XIV, 220 pp. 31 figs. Hardcover DM 158,–
ISBN 3-540-13019-5

Part D

1986. XI, 248 pp. 32 figs. Hardcover DM 204,–
ISBN 3-540-15555-4

Part E

1990. XII, 193 pp. 23 figs. 42 tabs. Hardcover
DM 128,– ISBN 3-540-51423-6

Volume 4

Air Pollution

Part A

1986. XI, 222 pp. 71 figs. Hardcover DM 182,–
ISBN 3-540-15041-2

Part B

1989. XI, 261 pp. 93 figs. 42 tabs. Hardcover
DM 198,– ISBN 3-540-50915-1

Volume 5

Water Pollution

Springer-Verlag Berlin
Heidelberg New York London
Paris Tokyo Hong Kong

Part A

Approx. 200 pp. 40 figs. 40 tabs. Hardcover
In preparation. ISBN 3-540-51599-2